BIOPSYCHOLOGY

BIOPSYCHOLOGY

*Scientific foundations
of human behavior*

Dr. Esteve Barrull Pons

1.ª edition: September 2023

© Esteve Barrull Pons, 2007

Cover design: Montserrat Prat Garriga
Elià Barrull Prat y Ariadna Barrull Prat
ISBN: 9798862170016
Independently published
Printed by Amazon

Table of contents

Preface

Introduction. The unitary paradigm

Part 1. Biophysics. Theoretical bases of behavioral phenomena
    Thermodynamics of life
    Information theory
    The behavior of living beings
    Consequences for the study of human behavior

Part 2. Biopsychology. Affection, health and well-being.
    The distinctive characteristics of the human species
    The nature of social support
    The exchange of social support
    Relationship between affective balance and health
    Affective relationships in today's society

A final consideration

Conclusions of the first reader

Editor's note

This book is based on a document called "Scientific foundations of human behavior" that I discovered on my father's computer after his passing in February 2022.

In this document, created in 2007, my father tried to summarize all of his scientific research into a guide to human behavior, so that my sister and I could survive in the midst of the intricate jungle of social relationships in which we all live.

Although he was unfinished, he had a clear structure and, for me, who knew his ideas and his discoveries well, the whole approach made obvious sense. Moved by curiosity and the nostalgia of our conversations, I spent several months searching through his files and notebooks for other texts that could serve to complete it. After gathering all of his writings, I based the book on the document I mentioned. I dedicated myself to completing the structure of the book, adding content to the incomplete chapters.

The result was the book you have in your hands. I could have kept this book in a drawer or made a few copies for friends and family. But I decided that I should publish it because, in addition to being a nice way to honor his memory, here you will find some ideas that I think are relevant.

Over his years of research, my father developed a coherent interpretation of human behavior based on thermodynamic systems theory and information theory. This particular point of view on what life is led him to discover the perception-action cycle that governs the behavior of animals. A simple and effective theory that powerfully explains animal behavior as a mechanism of adaptation to the entropy we perceive. This discovery reinforced his thermodynamic perspective on life and led him to develop the hypothesis that the exchange of affection, understood as a flow of energy, is closely linked to health and non-communicable diseases.

As you will see throughout the book, these discoveries allow us to understand human behavior, social relationships and their close connection with our health from a new perspective that is consistent with natural laws.

Elià Barrull Prat, 2023

Preface

"Knowledge does not take place"

Dear childrens Elià and Ariadna,

In this book, I aim to gather the majority of the results from over twenty years of studying and researching human behavior. The main purpose is to provide you with a guide to the scientific foundations of human behavior that can help you survive. In fact, I wouldn't trouble you if it weren't for the fact that I discovered something truly important:

The cause of success and failure, health and illness, lies in the kind of human relationships we establish throughout our lives.

We live as best we can, often unaware of the reasons behind our experiences. We are born, and after navigating the challenges of family upbringing, we venture into this world, making more or less friends, forming romantic relationships, pairing up and breaking apart, tolerating co-workers and superiors, neighbors, and the countless strangers who, for the most part, offer us little good but often the opposite. All of this occurs under the perpetual influence of inept parents and a family that, in general, only annoys and bores us. And all of it, ultimately, leads us to have children whom we struggle to guide despite our good intentions, watching them inherit our own failures and falter, gradually blending into the sea of mediocrity and failure that comprises us all.

Let me give you an example that illustrates the degree of failure in which we live. It is widely accepted that studies and training are a primary good for human beings. There can never be enough education and training; the more one possesses, the better for the individual.

It has always been said that in the past, the common people, peasants, and later the workers, lacked education because they were denied access to schools and universities due to their economic poverty. While the latter is true, I doubt it is the sole reason for their ignorance.

Today, access to public schools and universities is economically feasible for the vast majority of the population, let's say 90%. I mean that the costs associated with education can be covered by the vast majority of parents. However, we encounter truly eloquent facts:

1. A significant portion of students only access compulsory primary education, let's say 40%. This means they learn to read, write, and do basic arithmetic.
2. Another significant portion pursues university studies in fields other than natural sciences, such as languages, economics, psychology, sociology, history, and so on.
3. Only a small proportion pursues university studies in natural sciences, let's say 15%.

In other words, despite education being an intrinsic and fundamental asset in a person's development, the majority of the population does not reach a minimum level suitable for current times, despite having the necessary economic resources.

Today, the majority of the population is unaware of the fundamental laws of Mechanics and Thermodynamics, for example. They are unfamiliar with the basic principles of electricity, magnetism, or chemistry. They are unaware of the significant facts about the evolution of life on Earth and the laws that govern it. In other words, they are functionally illiterate despite knowing how to read, write, and do basic arithmetic. They live surrounded by technology, or even more so, they live thanks to technology and have no knowledge of how it works. The vast majority of the population is inept and incapable of solving the myriad of small problems that continually arise with the technological devices and appliances they use.

This illiteracy occurs despite full access to public education. Why do young people drop out of school or opt for non-scientific studies? It's not due to lack of money. The reason lies in parental ignorance and their cultural poverty. "If the child doesn't want to study, then let them work, as God intended," and problem solved.

The child doesn't want to study because it's difficult and very challenging, and if they have the slightest opportunity to choose another path, they don't hesitate to do so.

Now we take refuge in the excuse that "the child isn't cut out for studying, is foolish, or lacks the ability." Few lies are as cruel as this one. The one who truly "isn't cut out for it" is the parent who is unable to assist their child in continuing their studies, in overcoming difficulties and failures. I don't want to dwell on this topic now. Let's simply turn to the proven facts:

1. Per capita income and social assistance enable 90% of the population to access higher education in public institutions.
2. Only 15% of the population attains a sufficient level of scientific literacy.

As a result, 75% of the population remains on the sidelines, either not embarking on education or failing in it.

Another example that can help you grasp this tragedy is the so-called reality shows. Many progressive and educated individuals disparage these programs because they expose, even if just a bit, the harshness and misery of human beings in their everyday lives. In them, we witness how ordinary people routinely exhibit a high degree of misery and cruelty. Now, all you have to do is imagine that such behaviors occur under public recording and broadcast. How might this behavior be in private homes, without cameras or observers? I can assure you that horror movies are naive fantasies compared to the reality of our private and secretive lives.

So, just as we fail in the upbringing of our children, we fail in our relationships as couples and within our families, in friendship and companionship. We fail in our lives by not pursuing our true interests and callings. We fail in our health and well-being, developing multiple drug dependencies and psychosomatic illnesses.

I want to tell you that while we know how to go to the Moon, how to extract energy from coal or sunlight, how to implant a liver, how to stack 150 floors on top of each other, etc., etc., we have absolutely no 'damn' clue about how to live, love, and resolve the conflicts that daily flood our relationships with others.

What do I do with my father? What do I do with my partner? What do I do with my neighbor? What do I do with my mother? What do I do with my daughter? These are questions that each one has to answer for themselves, without assistance, on their own. Well, the only assistance is a few generic ethical-moral precepts of religious origin that, in most cases, are of very little use or the opinion of a friend who knows as much as you do.

Of course, the so-called 'experts' are as ignorant as oneself. I remember that one day I conducted the following test: In a doctoral psychology class that Pilar and I taught, I asked the new students (all psychology graduates) the following question: Could you tell us the five most important things you have learned throughout your career regarding human behavior?

The silence following the question was so thick you could cut it with a knife. Perhaps a minute passed without a word being uttered. I don't remember very well, but I think I even tried to ease the tension by saying something like, 'Well, you must have learned something about human beings in these five years of study.' In the end, one student, hesitatingly, said something along the lines of, 'Human beings are social creatures.' If it hadn't been for the context, I would have burst out laughing. Instead, I simply commented that my grandmother (I was thinking of my maternal grandmother María, but it could be any grandmother) also knew that and had never studied psychology (in fact, she didn't study anything in her life as she started working at the age of 12 in a factory). Of course, Pilar urged me never to repeat this experiment again. Well, so-called experts may be experts in something, but not in human behavior.

The result of so much ignorance, which persists for millennia, leads to the disaster of lives and human relationships. A disaster that we all try to hide with all our might. We always try to smile and say, 'Oh, everything is going very well.'

You might think I'm exaggerating, but I believe I fall far short, as I hope you'll come to see throughout this work. In fact, it's partly logical that it's hard to see the disaster in which we live because we can conceal it quite successfully thanks to money.

It is said that money doesn't buy happiness, but it does help, and in part, it is quite true. In fact, money doesn't buy happiness, but it helps to appear happy. We live in an extraordinary period in many respects, and one of them is the abundance of money. So much so that the poorest in our society live better than the wealthiest (kings, nobles) of earlier times before the Industrial Revolution. I'm referring to what they eat, wear, and the services they enjoy (medicine, education, transportation, security, etc.). In physical terms, it can be said that the poorest today consume much more energy (kW/h) than, for example, Julius Caesar did.

With money, we cover up, hide, dress our sorrows and frustrations, but we remain just as miserable and ignorant. I see myself, and all other humans, as wretched kings, dressed in elegant clothes, surrounded by marvelous contrivances, seated before exquisite feasts, perfumed with a thousand fragrances, numbed by abundance, pretending as if we were the happiest in the world. Yet deep inside, we are filled with fear and disorientation, sick, desperate, corroded and rusted, waiting for the slightest opportunity to rid ourselves of so much bitterness and despair.

That's why it's difficult to see the harshness of our lives. Who would believe that the woman dressed in such expensive clothes, smiling generously and exuding exuberance, is deeply consumed by fear? Who would believe that this gentleman feels profoundly alone, despite the wonderful family that adores him and the success he achieves at work? Almost no one.

What I'm going to try to teach you in this book is to see that behind this magnificent appearance lies not only a person who suffers but also a person who can make you suffer terribly if you allow it, if you fall into their grandiose web of seduction and into your own naivety of 'do good and don't look at who.'

It's only necessary to take a look at the statistics on marriages/separations, domestic violence, school failure, delinquency, drug addiction, disputes, accidents, abortions, and diseases to get an idea that beneath the appearance of a welfare society, luxury, and happiness, lies a suffering society, and consequently, an ignorant and ultimately, cruel one.

However, as we are accustomed to living in a disastrous way, and it's a common affliction of all mortals ('misery loves company'), we accept the situation as normal and even inevitable.

I aspire to impart knowledge to you that will help you climb out of this pit of ignorance and enable you to organize your lives and relationships with more than just ethical-moral precepts and good intentions. Knowledge that will empower you to take control of your life, to make informed decisions about what is best for you in each relationship, with respect to each person who comes into contact with you. This will allow you to make the most of what life offers, to not waste resources or opportunities, to avoid falling into prisons and traps that would destroy you without you knowing what to do. It will allow you to develop as individuals both deeply and extensively.

Now then, there are no magic formulas, only the effort of knowledge and study. Only with knowledge can you overcome a problem, and as you must already know, knowledge always requires a lot of effort.

Allow me to provide some context on how I arrived at this result in my research. As you know, I graduated in psychology after studying architecture for four years. I have no doubt that this change is due to my relationship with Priest Lluís Munté, who guided me to confront my emotions and taught me the value of solitude between the ages of 16 and 20. I decided to study architecture because it provided me with a broad range of education, from the sciences (primarily mathematics and mechanics) to the arts (form, color, and design). However, I soon realized that I was truly more drawn to the study of emotions and human behavior, so I began studying psychology.

When I was 25 years old, during a 2-month trip to Ecuador, I decided that I wanted to dedicate my life to the study and research of human behavior. I didn't want to live a standard life; I wanted to embrace the risk of an adventure, and I realized that research was my personal adventure. Despite living in a society that often undervalues the importance of science, where the famous idea is "let's enjoy what others invent," and where people believe that dedicating time to research is foolish, I organized my life in order to minimize expenses and economic work. This allowed me to have a significant amount of free time to carry out my plan.

You also know that I worked as a therapist and as a group psychology professor at the university for 20 years until 2004, not as a means to earn money, as I never did, but as necessary activities to complement my studies and research.

Since I began studying psychology, I was deeply disillusioned by the complete lack of scientific rigor. Psychology seemed to me like a disorganized collection of speculative pseudo-theories, more dedicated to explaining "what human behavior should be" rather than "what human behavior actually is." Classes and textbooks were filled with beautiful phrases, good proposals, and ambiguous and neutral language, where nothing was definitively established. Psychologists appeared to me as individuals capable of saying nothing with many words.

From the very beginning, I felt that psychology needed a radical change, one that involved studying human behavior as a natural phenomenon, especially as a physical phenomenon. You know that physics is the foundation for understanding all phenomena, and I was convinced that we wouldn't be able to comprehend human behavior until we could understand it from a physics perspective, at least in a general and abstract manner.

As a student, I was fortunate to meet Professor Pilar González, who taught Group Dynamics. She argued that human behavior was a biological phenomenon and that our behavior was truly driven by our emotions rather than our rationality. These ideas caught my attention, and I tried to become her disciple. Fortunately, she also took an interest in me, and gradually, a mentor-disciple relationship developed between us.

In addition to being a professor at the university, she worked as a therapist two afternoons a week as a way to stay in close contact with empirical facts. She allowed me to work with her and taught me everything she knew. In particular, she helped me understand how human behavior should be understood solely as a biological phenomenon, which was the main thesis of her professor, V. Wukmir. Those were great years for me because of the close contact with real issues of human behavior. I enjoyed intense theoretical discussions with her that led me to develop a very complex understanding of human behavior. Without a doubt, she has been the greatest influence on my scientific development.

Simultaneously, I was seeking a way to treat human behavior as a physical phenomenon, primarily through mechanics and thermodynamics. I even explored the possibilities of quantum mechanics[1] and relativity[2], although I understood that they couldn't function as the primary reference for addressing human behavior. Another significant aspect was the study of information theory because I was convinced that it had to play a central role in this project.

---

[1] For example, how a particle was trapped by a potential well depending on the particle's energy suggested to me the way stores could capture customers walking down the street at different times. Furthermore, it seemed suggestive to describe a relationship as two particles trapped by their respective potentials.

[2] I explored how relativity could take into account the relative perception of time. It is known that we sometimes perceive time as slow and other times as fast. I explored how this could be related to the intensity of behavior, meaning that as you move more, you perceive time as faster (time dilates), and vice versa.

To maintain discipline, I committed to a research project that would serve as my doctoral research and could centralize most of my various research activities. Without rushing, but dedicating a lot of time to it, I began an investigation that represented my initial ideas of what the psychology of the future should be. My premise was to study human behavior from a physical perspective (but not yet as a biological phenomenon).

Under Pilar's guidance, I was able to find theoretical support in the principle of psychophysical isomorphism developed by Werteimer and Köhler, the founders of Gestalt psychology. In summary, this principle states that the physical order of behavior corresponds (is equal) to the physical order of the neural activities responsible for that behavior and to the subjective order of our related experience.

Applying this principle, I established that the order of emotional states (neural activities) should correspond to the order of vocal activity during speech. In other words, we could determine how positive or negative a speaker's emotions must be by examining the organization or disorganization of their speech. I knew that speech is always organized in acoustic spectra (20-20000 Hz) because it reveals only the activity of the vocal cords for producing vowels and consonants. However, there had been no study on syllabic activity, i.e., the frequency range used to produce syllables (2-20 Hz), which corresponds to the activity of the jaw, tongue, and breathing during speech. Considering a syllable as a packet of acoustic energy, the question boiled down to calculating the Fourier transform of a sequence of these packets and then calculating their entropy.

After several years of building the necessary tools to conduct such measurements, I couldn't believe what I was seeing when I collected the initial results. It didn't matter who the speaker was, what time it was, or what situation they were in; I consistently found the same results. The syllabic spectra took on the form of a Boltzmann distribution in a very short time, let's say, a minute. As you know, this distribution corresponds to the maximum entropy for variables that range from zero to infinity, as is the case with a spectrum.

After thoroughly verifying every element of my computational tools, I had to acknowledge this unexpected result, which was followed by a year of depression.

However, during the liberation war of Kuwait, I found a theoretical explanation. I was thinking that the syllabic behavior seemed as if the speakers were speaking in a foreign language to me, a completely incomprehensible language. Listening to news about the war preparations, I remember pondering the problem of concealing communications from the enemy through encoding tools. I quickly glanced at my information theory books to find some clues in the chapters dedicated to coding. What a joy it was when I discovered Shannon's main coding theorem in the book!

In summary, this theorem states that to achieve the most efficient communication, messages should be encoded in such a way that the transferred signal must have maximum entropy, precisely what I found in my research. The price to pay for such efficiency is that the encoder must be different for each different source of information, meaning the encoder must adapt to each specific source. With this theorem in hand, I was able to solve my research in the following way:

1) There must be a neural encoding system in our brain for the transmission of emotional information through syllabic activity.

2) The existence of a highly efficient encoder only makes sense if the information transmitted through syllabic signals is very high.

3) Because syllabic encoders are the most efficient, human misunderstanding in verbal communication is a result of decoding errors due to each person having their own specific and different syllabic encoder. Only through a close relationship can these errors be reduced.

To visualize these results, you can imagine what happens when a group of people is instructed to speak in unison, reading the same text, for example. This means they must synchronize their syllabic production, in other words, they have to suppress their personal encoders. To achieve this, rhythm (order) must be introduced into their syllabic production to facilitate matching. In such a case, the speech loses emotional information and becomes less human and more robotic or automatic.

I delved into my doctoral thesis research because it was the drive that allowed me to discover the general way to understand human behavior as a natural phenomenon.

Right after presenting my doctoral research thesis, I decided to revisit the study of information theory as a way to express my gratitude for its pivotal role in my research. I studied preliminary information theory through various books, some of which were easy for me, while others were more challenging. One of the most challenging books was Shannon's, but since it was in his book where I found the solution to my thesis problem, I decided to start with it. Very soon, I realized that I had been making two significant errors, one concerning the concept of information and the other regarding the measurement of the probability of an event.

In some 'easier' books, I learned that the information conveyed by a symbol within a message was simply the logarithm of the reciprocal of its probability, meaning,

$$I(x) \equiv - \log p(x) \qquad (1)$$

With this idea in mind, I couldn't find a way to apply this concept to human behavior. The major drawback of (1) was that, in my experience, the same stimuli conveyed different amounts of information to different receivers[3]. The same news makes one person happy and another sad. So, I couldn't find a way to apply this definition to human behavior.

But in Shannon's book, I found another concept of information, which was defined to account for the interference of noise during transmissions. In such cases, the received symbols, y, did not

---

[3] In fact, the information defined in (1) is not the "transmitted" information but rather the "carried" information by the symbol, or the maximum information it can convey. Of course, without noise interference, it is the transmitted information, but that's only an ideal case. In practice, it represents the maximum information a symbol can transmit.

necessarily match the transmitted symbols, x. So, the information conveyed by the received symbols y about the transmission of the symbols x is given by

$$I(y, x) = H(x) - H(x/y) \qquad (2)$$

That is to say, the uncertainty about the emitted symbols is reduced by the received symbols y, and about the emitted symbols x.

Such a definition suggested that information was the reduction of uncertainty in the receiver. Taking into account that different receivers could be defined with different noise interferences, the same message would convey a different amount of information to different receivers. This fulfills an appropriate definition of information to be applied to human behavior. With this idea in mind, I only needed to step out of information theory and consider humans as sources and receivers of real information.[4]

Then I realized that I could formulate a general definition of information as a subjective experience (emotion) that actually worked in the phenomena of human behavior. I was able to define information as the experience of reducing uncertainty.[5] In other words, humans experience information as the emotion of reduced uncertainty. Every time we experience an increase in peace, stillness, tranquility, freedom, happiness, security, hope, etc., or every time we experience a decrease in anxiety, restlessness, violence, despair, etc., it means we are experiencing the reception of information because all these emotions are directly related to or are specific cases of the emotion of

---

[4] You must remember that information theory is a mathematical theory that doesn't concern itself with empirical facts. It is a tautological theory that doesn't say anything about the real world.

[5] In fact, it's just Shannon's definition, but I suspect he didn't want to extend his work beyond where he felt comfortable, which is the realm of mathematics. I believe he was aware of this, as indicated by his explicit warning at the beginning of his work: "Messages often have meaning; that is, they refer to or are correlated according to some system with certain physical or conceptual entities. These semantic aspects of communication are irrelevant to the engineering problem." (Shannon, 1948, p.1)

experiencing reduced uncertainty. So, positive emotions can be directly related to the reduction of uncertainty (information gain), while negative emotions are directly related to an increase in uncertainty (information loss)[6].

Therefore, humans avoid uncertainty (negative emotions), which is an evident fact, and it means that humans desire information (positive emotions), something that is also an evident fact! I learned from Pilar that humans are driven by their emotions, which I could continually confirm through my experience. However, now I found an expanded view of this fact. I could rephrase such a statement by saying that we, humans, are subjectively driven by our need for information or our aversion to uncertainty.

However, this small step would have been inconsequential if I hadn't found any connection to the objective world of human behavior. Fortunately, I was able to find it thanks to the identity between Shannon's uncertainty and the Boltzmann entropy function.

My greatest surprise was realizing that information, as defined by Shannon, was intimately related to Boltzmann's entropy through the concept of uncertainty! Shannon discovered that the best function to define uncertainty was Boltzmann's entropy function.

Boltzmann defined entropy as

$$ S \equiv - \; k \cdot H_{Boltzmann} \equiv - \; k \int p(v) \cdot \ln p(v) \qquad (3) $$

where $p(v)$ was the velocity distribution of particles in a system, and k was the so-called "Boltzmann constant." Then, Shannon, taking the discrete formulation of Boltzmann's function H, defined uncertainty as

---

[6]If you feel uncomfortable with the words "information loss" because you remember from your studies that information can only be positive or zero but never negative, please wait a bit because I will address this issue later on.

$$H_{Shannon} \equiv - H_{Boltzmann} \equiv - \sum p(x) \cdot log\, p(x) \qquad (4)$$

where *p(x)* was the probability distribution of events *x*. The small differences were only a matter of convenience, and the function *H* was exactly the same in both cases.

So I saw clearly that one of the main enigmas could be resolved. The behavioral sciences are the only ones to have two viewpoints of the same facts: objective and subjective experiences. In science, we usually deal with only objective experiences, meaning the experience of things from the outside. However, we can also experience our own behavior from the inside, or what is called subjective experience. Therefore, a true science of human behavior must be able to describe the same facts from both objective and subjective experiences coherently. Thanks to this mathematical identity, I found the key to managing both objective and subjective experiences with the same theoretical background. The path was simply to translate statements from the objective world of statistical mechanics into the subjective world through information theory.

Allow me to provide a simple example. Statistical mechanics states that "if an isolated system is not in its thermodynamic equilibrium, its entropy will increase." This statement can be applied to human subjective experience if we translate it into the language of information theory: "if a person isolates themselves, they will lose information, meaning they will experience an increase in uncertainty (negative emotions)." This statement is the result of translating 'entropy will increase' into 'will lose information.' If uncertainty increases, information decreases, and negative emotions are experienced.

Another example: statistical mechanics states that a system can only become more ordered through interaction with its environment. Similarly, we can translate it as: a person can only gain information by interacting with their environment.

In other words, there is only one world of physical phenomena of human behavior explained by a single theoretical background, which is given by statistical mechanics (thermodynamics). But this world can be expressed as it is objectively observed (from the outside we observe the facts) or as it is subjectively experienced (from the inside we experience emotions). In the example above, one can predict an increase in entropy and also an increase in negative emotions. Both predictions refer to the same physical facts but are observed (experienced) from two different viewpoints (external and internal). Later on, we will see how the physical foundations of this remarkable relationship between these two viewpoints can be understood through the Norwich law.

The main enigma was about to be solved. I was convinced that human psychology had to deal not only with observed facts but also with subjective experiences. It was one of the disadvantages of psychology. Behavioral scientists studying other species don't need to explain the emotions of the subjects, only what they observe from the outside, that is, objectively. But I thought it was necessary to understand human behavior by comprehending human emotional experiences. In this way, the bridge that enables the connection between these two worlds became possible.

Very soon, I found the idea from Schrödinger about considering living beings as entities that feed on negentropy[7] to be suitable and highly valuable, which I read many years ago. Consequently, human behavior would consist of a continuous search for information, a continuous attempt to reduce uncertainty, from a subjective point of view. In other words, we are continuously trying to satisfy our curiosity, experiencing ourselves as beings that feed on information. Exactly the same thing happens from an objective point of view; human behavior consists of a continuous search for order, a continuous

---

[7] The concept of negentropy was coined by Léon Brillouin, a 20th-century French physicist, based on Schrödinger's idea of negative entropy as presented in his book "What is Life?" Brillouin's objective in using this term was to demonstrate the similarities between the uncertainty defined by Shannon and Boltzmann's entropy. (Editor's Note)

attempt to reduce entropy. Through the identity between subjective uncertainty and objective entropy, I established the proper path to understand the entirety of human behavior phenomena through the natural sciences.

But to achieve this, I needed to overcome a second error. Information theory and statistical mechanics are probabilistic theories. Statistical mechanics is an empirical theory. Its truth is derived from empirical research, and the way it measures probability is clearly defined: the probability of an event is defined as the number of different ways in which particles can be microscopically organized to yield the same macroscopic event, relative to the total number of configurations that these particles can achieve. On the other hand, information theory is a mathematical theory, so it doesn't relate to empirical facts. In such a case, the probability of an event is not defined in relation to the physical world but simply as the number of occurrences of an event relative to the total occurrences of the set of all considered events.

Due to my years of studying traditional psychology and its "scientific" research methods, I didn't realize that I had acquired a mistaken concept of probability. Psychologists measure the probability of events mathematically, without considering the physical nature of events. For example, the probability of having a car accident within a month is measured by the number of car accidents divided by the total number of car trips within that month. As you know, this probability is always very low compared to the probability of having a safe car trip.

However, the inconsistency became evident when I realized that from the perspective of statistical mechanics, the probability of having a car accident is enormously higher than the probability of having a safe car trip. The order and discipline of cars on a road are something very improbable throughout the universe. It is very unusual for pieces of matter to be arranged as cars moving orderly on a road.[8]

---

[8] Of course, this movement of cars is caused by the living beings that drive them, and living beings are also very improbable events.

In summary, statistical mechanics can be applied to understand subjective experience through information theory if the probability of events is measured in the manner of statistical mechanics. Continuing with the previous example, a car accident is a source of very high subjective uncertainty (negative emotions) because it is highly probable (thermodynamically). If we ignore the physical nature of a car accident, i.e., the destruction of a lot of material order, and only consider its frequency, not relative to the universe but only to car trips, we will conclude a very low probability for a car accident, and then we must consider it as a source of information or positive emotions, which clearly goes against our experience.

I'm sure you have many questions as you read these paragraphs, but you must understand that here I am only trying to explain the origins of my discoveries in a very brief way. Throughout this book, I will properly explain each of these topics, and I hope to address most of your doubts.

After these initial steps, between 1993 and 1994, I incorporated biophysics, Dawkins' cultural approach, Norwich's discovery of sensory neuron behavior, Wukmir's concept of emotion, and modern ecology of communities to draw a complete theoretical foundation and understand human behavior as a natural phenomenon, both from the subjective and objective perspectives. I also discovered M. Rothschild's book "Bionomics," which showed me that I was not alone in pursuing this path. It was the most compulsive time of my life. I felt like I was fitting together a complex puzzle that clearly depicted a coherent image of real human behavior.

And that's why I remember crying deeply when I first realized the profound significance of my discovery. I didn't cry out of happiness but out of frustration, disillusionment, and helplessness because I understood that any idea of objective human freedom must be abandoned. That goodness and cruelty were intimately mixed by the determination of natural laws. No form of idealism had a place in the modern science of human behavior. And I also remember that when my teacher Pilar first understood the scope of my discoveries, she also cried.

This is the price we must pay for knowing the truth about our behavior, and as I have found many times, people normally cannot bear it. I remember many students asking to leave our classes (both Pilar's and mine) because they felt ill.

Thousands of years believing in human superiority, in human privilege to decide our destiny, in human abilities to prevent cruelty and destruction in the world, and this change is so contrary to our deepest desires that it turns our stomachs. I recognize that it is not an easy change, no matter how many benefits it may bring. It will take hundreds of years for human culture to accept such results.

However, you should not think that this perspective leads to destruction. On the contrary, the fact that natural laws deny the absence of cruelty and suffering provides us, and those who understand human behavior through them, with powerful tools to intervene in the world to reduce cruelty and suffering. In other words, by assuming cruelty and suffering as natural aspects of our life, we are in a position to minimize them. The history of science is full of similar situations. By assuming that the Earth was not at the center of the universe, humanity was able to travel to the Moon. Of course, we cannot put the Earth at the center of the universe, but we can do many positive things thanks to modern astronomy.

I'm sure the question that comes to your mind is: if it's such a useful discovery, why haven't you published it in a scientific journal[9] or shared it in scientific circles, instead of writing it for us?

I did not talk to you about my scientific activities for several reasons. First, because my ideas were firmly rejected wherever I presented them: students and professors at the university, patients in my work as a therapist, journal editors, and the general public at conferences and seminars. Only a few friends made an effort to listen

---

[9] Of course, I have tried to share this work, but with very negative results. And since I don't want to be a martyr of science, I decided to postpone the issue and leave it for future times and generations. That's why I'm writing to you. I sincerely believe that it can be very useful to you if you are able to delve into the subject.

to me and understand to the best of their abilities, for which I am grateful. Because I did not achieve any social success as a result of my scientific activities, there were no occasions to discuss them with you.

Another reason was that I did not want to influence your future more than necessary, especially not to convey the idea that natural sciences were better activities than others, considering that I pushed you to attain a moderate level of a natural sciences bachelor's degree.

So, what you have in your hands is the result of my scientific activities related to the study of human behavior. I dedicated the most important part of my life to this subject, from the age of 20 to 47. After that, I stopped this research to preserve the health of my brain. However, I decided to write a book to present the results of my research. This book is aimed at you, my children, because I hope that you will be curious about your father's activities and also because you will be able to understand it due to your special scientific education. Of course, I do not reject the possibility that some other reader might be interested in this book, but I acknowledge that the odds are very low. And if another reader becomes interested and reaches the end of this book with benefit, they will also be my child.

One of the reasons my ideas were systematically rejected was that they were the result of pushing current scientific knowledge to its limits and also due to a deep understanding of emotional experiences. Only individuals with a broad scientific background and no fear of dealing with human emotional experiences will be able to comprehend what is written on these pages. Unfortunately, current learning policies tend to produce specialized scientists who have deep knowledge of a small area of phenomena but are unaware of and even disinterested in many other phenomena, especially emotional ones. Unfortunately, our chronic rejection of our emotions in favor of rationality prevents this book from being understood as well. In fact, I discovered many years ago that our rejection of emotions is because we recognize that they cannot be voluntarily manipulated as reasoning can be. So, accepting that emotions govern our behavior would imply that we cannot voluntarily decide our behavior, which is completely unacceptable to the population. They prefer to believe that they are ruled by their reasoning, which is easy to manipulate.

The main reason I wrote these pages was not to lead you into the study of human behavior and follow in your father's footsteps. The reason is to provide you with essential knowledge about human behavior that can help you succeed in your life. By studying these pages, you will understand better than anyone else the main rules that govern your behavior and that of the people around us. My hope is that this book can help you succeed in your social relationships throughout your life. More than that, it could help you maintain a healthy life by avoiding many dangerous human relationships. I hope you have enough time and resources to study this book. Although my contemporaries vehemently rejected my results, I am sure they are good and truly accurate, meaning they align with the facts of human behavior.

Of course, you should read this book critically, as is customary in the natural sciences. The truth lies in the facts, not in a book. There are two main sources of disagreement with these pages. First, it may evoke negative emotions in you, something that is likely to happen. However strong they may be, you should try to remain calm and remember that science pursues what the facts are, rather than what they should be to satisfy our desires. Recall that historically, science has revealed facts contrary to human wishes, leading to strong and profound conflicts among people due to their different emotions.

Second, you may recognize that something is indeed an error. I am sure that will happen as well. Although I am convinced that the overall plan is correct, I know that I have made many errors in these pages due to the extensive range of phenomena discussed. In some of these cases, I am also sure you will know the correct answer.

Regarding rejecting the entire plan, I would ask for deep reflection on whether hidden negative emotions are motivating you to do so or not. Since I arrived at this theoretical plan, I have been waiting for the day when I could find some major error that would lead me to reject this project. Unfortunately, that day has not come yet. Perhaps you can find it and even achieve a better understanding of human behavior.

It is true that I may be wrong, but that's what you and any other reader are here for. It's up to you to judge whether I am right in the light of your knowledge. I only (as if that were not enough) ask for the effort not to reject these ideas without good reason or simply because you don't like them. If you choose to reject them, let it be because you can properly reason where I have made an error. These ideas are not the result of acts of faith but rather of argumentation and the integration of knowledge from various scientific disciplines.

The fact that I compelled you to achieve a deep and broad knowledge of natural science was related to my research. Although I primarily justified my imposition by saying that science encompasses everything around us, mainly through technological advancements, the truth is that the most important reason was my conviction that the natural sciences are the key to understanding human behavior. Knowing that human behavior, especially human relationships, is the primary factor in success or failure in life, I believe that my best legacy would be to give you the best of myself, which is this guide to human behavior.

Esteve Barrull, 2007

Introduction: The Unitary Paradigm

The entire book can be considered an approach to understanding human behavior, and especially its close relationships with human health and well-being from a new perspective: considering humans simply as living beings, that is, studying human behavior in the same way we could study that of ants or elephants.

Traditionally, the study of human behavior is considered entirely different from the study of the behavior of other species because it is believed that human behavior does not follow the same rules as that of other species.

I call this new perspective the "unitary paradigm" in contrast to the traditional view, which I call the "dualistic paradigm."

The unitary paradigm

Before we begin our study of human behavior, it is necessary to explain the particular perspective we will use to carry it out. Our viewpoint is to consider that

we humans are just living beings,

and thus we must be studied to achieve a scientific understanding of our behavior. From a scientific standpoint, we view human beings as they are, that is, only as animals. No matter how many differences there may be between a human being and any other species.

Once the biological nature of the human being is accepted, we can even delve a little further into human nature. The modern biological paradigm states that the phenomenon of life is only a physical phenomenon governed by known physical laws (Crick, 1958; Dawkins, 1976; Dennett, 1991; Monod, 1970). Microchemical knowledge of living processes ensures this premise, and the modern thermodynamic approach to living phenomena promises great advances in the coming decades. Therefore, we could rephrase the unitary paradigm as:

All human phenomena are the result of the properties of matter.

The unitary paradigm emphasizes materialism as the proper way to understand human behavior. There are only atoms and their interactions. Everything arises from them. Of course, we cannot take into account the atoms involved in human behavior, nor the molecules, cells, tissues... or even make any useful measurements. We are extremely far from being able to analyze human behavior in a manner similar to how physics analyzes the efficiency of a car engine. But instead of speculating freely, allowing any kind of claim and ignoring the laws of physics, we restrict our analysis of human behavior to work only with statements that are compatible with known physical laws.

The dualist paradigm

The traditional viewpoint separates human phenomena into two parts: natural phenomena and spiritual ones.

Humans have characteristics
that are not of a physical nature but spiritual.

Since its philosophical development in the 17th century (Descartes, 1641), the dualistic paradigm defines anything that falls outside the realm of natural laws or cannot be explained by natural laws as having a spiritual nature. Phenomena like language, thought, science, or religion are seen as beyond the scope of natural laws and are attributed to a spiritual nature. It is believed that these phenomena cannot be explained by physical laws and that we need to find a different set of laws to understand them.

This traditional attitude likely stems from the fear of acknowledging that humans are simple animals. In part, this fear is justified by past negative experiences. Philosophies like social Darwinism contributed to associating Darwin with inhumane practices (Dennett, 1995; Gould, 1981; Hofstadter, 1944). However, another part of this fear comes from ignorance about how science works.

Many people think that considering humans as animals can lead to a denial of the value of solidarity, charity, helping the weaker, affection, or human compassion (Kass, 1985; Lewis, 1943; Scruton, 2000). In essence, they believe it promotes violence, cruelty, and terror. However, nothing could be further from the meaning of the unitary paradigm.

The unitary paradigm does not deny the value of solidarity, charity, affection, or compassion in human behavior. On the contrary, as we will see in Part II of this book, they are crucial for our survival. What the unitary paradigm asserts about these things is that they result from natural laws and not hidden spiritual laws (Darwin, 1871; Dawkins, 1976). According to this approach, the fact that humans are the most solidary, affectionate, and charitable species in the history of life on Earth is simply a result of natural laws.

Solidarity, charity, and affection can be constantly observed in our behavior. This is an empirical fact that no observer can deny. The difference between the dualistic paradigm and the unitary paradigm lies in their responses to the question: What is the nature of these phenomena? The dualistic paradigm answers that they are of a spiritual nature, while the unitary paradigm answers that they are of a biological nature.

Later, you will realize how, by studying human behavior from the unitary paradigm, we can understand that solidarity, charity, and affection are the most important characteristics of human beings at a level that no other approach can reach.

The unitary paradigm is not a choice

Maintaining the idea that humans have a single biological nature is not merely a whim based on romantic love for the natural world, plants, and animals. It is simply a necessary consequence in light of current scientific knowledge. In fact, this is not a desired thought, at least not for me, but rather the imposition of facts that modern science has revealed about our nature.

If scientific ideas were a matter of free choice, I would have chosen the traditional viewpoint that suggests that human nature is, at least in part, free from the domain of natural laws. I would prefer a world where my behavior was a matter of free will, where I could be the master of my destiny. Personally, I find this approach more pleasant than the unitary paradigm, but unfortunately, our current knowledge does not allow me to support it.

In fact, there are many sources of data and scientific knowledge that support the viewpoint of the unitary nature of all living beings, including humans.

Millions of fossils, collected by paleontologists over the past centuries from all corners of the Earth, fill hundreds of thousands of boxes and drawers in museums, universities, and private collections worldwide. All these fossils have helped paleontologists reconstruct the history of life on Earth. Despite significant gaps in the chronology, there is unanimous consensus among paleontologists: all fossilized bones show that every species, including humans, descends from other species in such a way that all species have a common ancestor (Darwin, 1859; Gould, 1989; Johanson & Edey, 1990; Leaky, 1994). They are all related to each other like branches of a tree, and the history of life can be compared to the history of a small garden.

In a small, open garden, different plants and trees began to sprout. There was enough soil and light for all of them to grow, as they were very small at first. However, as they grew, they started to compete for soil and light. Their roots and branches were in close contact with those of their neighbors, and many of them stopped growing. After a while, only a few plants and trees remained alive, but as they continued to grow, constant competition eventually left only one tree alive. This tree is now very large, with hundreds of branches. Throughout its long life, it has lost many branches, some of them very large, brought down by significant events. It has survived storms, hurricanes, fires, and competition from other trees. Now, it stands alone in the yard, large and old, facing the challenges of the future.

This is the story of the evolution of life on Earth. New species (branches) arise from the old ones. Simultaneously, many species go extinct, leaving opportunities for other species to evolve. But the most important fact that concerns us here is that all species are related by a common ancestor. So, how is it possible that some living beings, which are completely subject to natural laws, have had descendants with a spiritual nature that escapes natural laws? From a scientific perspective, this question has no answer. Things that are governed by natural laws cannot produce something that is not subject to those laws.

Fossils show us that humans are descendants of ancient primates very similar to modern chimpanzees. However, according to the dualistic paradigm, while chimpanzees do not have a spiritual nature, humans do. This raises the question: which primate was the first to give birth to a baby with spiritual functions? The question itself reveals its absurdity. Only outside of scientific discussion can one continue with this topic.

The facts of evolution, unanimously proven and accepted, demonstrate by themselves that humans can only be governed by natural laws, in exactly the same way as any other species. Human behavior must be studied from this perspective, and we must reject the idea that there are spiritual laws or supernatural forces that influence our behavior. The history of the sciences of human behavior confirms that this approach is correct, and any attempt to explain it through spiritual properties is doomed to failure (Dawkins, 2006; Harris, 2004).

However, not only fossils are a clear source of data to support the unitary paradigm. All the data that come from anatomy and physiology support, without exception, the idea that humans are made of the same material and the same type of design as other species. From the perspective of anatomy and physiology, humans are classified as

Kingdom	Animal
Phylum	Chordata
Subphylum	Gnathostomata
Class	Mammalia
Order	Primates
Family	Hominidae
Gender	Homo
Especie	Homo sapiens

This means that nothing strange or unclassifiable has been found in our anatomy and physiology. There is no organ, bone, tissue, or type of cell that cannot be found in many other species. Our bodily systems, such as the circulatory system, respiratory system, and nervous system, are similar to those of other mammals. All this evidence points to the conclusion that humans are just another animal species, governed by the same natural laws as any other living being.

Even a child can understand that if humans eat only plants and animals, they must be made of pieces of plants and animals and nothing more. How is it possible for something made only of material parts, completely subject to natural laws, to behave, at least in part, independently of such laws? The logic of current science has no answer.

If we adopt the viewpoint of embryology, we will face another interesting aspect of the paradox. You know that all sexually reproducing living beings come from a single female cell that has been fertilized by a male cell (Gilbert, 2000; Wolpert et al., 2002). As individuals, we are initially just a single cell, nothing more, and only through intense exchange of material with our environment can this cell replicate itself and produce a multicellular organism like us. But are there spiritual faculties in the fertilized human egg? If the answer is yes, the question is where they are. If the answer is no, then the question is when and how these new faculties appear during embryonic development.

Finally, genetic research in the last half-century has brought surprising confirmation of evolution and the inner core of living beings. Each of today's living beings has within the nucleus of all its cells a set of macromolecules that store a complete program to govern its growth and physiology. This set of macromolecules (genes) is like a large book that explains all the details to form the individual and direct their behavior.

This book is written using only four symbols (C, T, U, G), which are small molecules, and a separator. The large sequences of these four symbols explain each of the individual's physiological characteristics (Watson & Crick, 1968). For example, a particular sequence of C, T, U, and G can determine a person's eye color, while another sequence can determine their tendency to be aggressive or empathetic. In this way, it is coherent to understand that living beings are governed by natural laws, as they are the result of the physicochemical interaction of large sequences of molecules with their environment.

In addition to demonstrating that humans are the result of the program stored in their genes, genetics has shown that 98.7% of human DNA nucleotides are identical to those of present-day chimpanzees. If we consider only the small fraction of human DNA that differs from chimpanzee DNA, the question that arises is: how can a different arrangement of these four macromolecules along this fraction of DNA be responsible for spiritual functions? Again, we are

41

faced with a meaningless question. It is absurd to suppose that a DNA sequence, which is entirely governed by natural laws, can produce spiritual phenomena that are not governed by natural laws.

Although each of the aforementioned sources of data is capable of establishing on its own the physical and biological nature of humans, the most conclusive data comes from neuroscience. Research in neuroscience over the last two decades has demonstrated that all faculties considered spiritual are a direct result of the electrochemical activity of millions of neurons. Thought, idea, emotion, or language are not the activity of immaterial processes within us but the physical activity of cells in our brain, which can be measured and observed thanks to new technologies (Sacks, 1973; Sacks, 1985; Damasio, 1994; Pinker 1994; Ramachandran & Blakeslee, 1999).

Traditionally, the nature of something has been attributed as either physical or spiritual based on the observation of changes. Something has been considered physical if it can be directly associated with observable changes. For example, something is considered physical if it can be touched, meaning we can experience a force opposing our own force upon it.

When we think while driving home, nothing seems to move or change in our perception directly related to the act of thinking. It appears that 'thinking' happens without apparent physical effort. This is why phenomena that primarily involve brain activity have often been the last to be accepted as physical in nature.

But modern neuroscience has come to our aid. Using various technologies, we can observe what things change and move when we think. And we are amazed to see that millions of neurons change rapidly when we experience a thought. So, what is the nature of a thought? Neuroscience provides us with solid evidence, direct data, that allows us to answer without hesitation that any thought is of a physical nature, like our hands, dogs, and stars.

In summary, evidence from paleontology, anatomy, physiology, genetics, and neuroscience all point to the conclusion that humans are purely physical beings governed by natural laws. Nothing in our existence is of a spiritual nature, and we only obey natural laws. But remember that this conclusion belongs only to the scientific realm. It only holds value within scientific work, and science is far from knowing the truth of our universe.

I am the first to be sure that there must be other types of realities far beyond physical matter and energy, beyond what we know scientifically. The more one knows about the world from a scientific perspective, the more convinced one becomes that science only covers a very small fraction of the world. The unitary paradigm is a scientific paradigm, so it is only useful within the scientific realm, and it would be a great mistake to try to expand it outside scientific discussions. The main lesson that science provides us is to feel a deep respect for the unknown world.

The sciences of human behavior

The unitary paradigm impacts all the sciences of human behavior, such as medicine, history, geography, anthropology, linguistics, pedagogy, economics, law, and sociology, among many others. Additionally, even though they cannot be directly considered as behavioral sciences, disciplines like architecture and engineering are also closely related to behavior and are therefore affected by it.

You may have noticed that your father omitted psychology from the list above. The reason for not including it is that psychology should be considered a subdiscipline of medicine rather than an independent discipline.

Allow me to reason this statement concisely. Regardless of what psychologists may claim, if you ask which discipline studies the relationships between human behavior and health and well-being, you can only give one answer: psychology. No other discipline is more interested in human health and well-being than psychology. However, because current psychology is far from achieving the status of a natural science, it is very difficult for medical professionals to accept it as part of medicine, which is a true natural science. Therefore, when psychology is able to work from the same perspective as medicine, it will be easily accepted as part of it.

Returning to the unitary paradigm and its dominion over all the sciences of human behavior, you must understand that the first consequence is that all these sciences should be considered subdisciplines of biology, just as biology should be considered a subdiscipline of physics, which is the main and fundamental discipline for all explanation and scientific knowledge of the universe.

Of course, all the disciplines of human behavior are far from implementing this program, although there are some indications in that direction. For example, some anthropologists attempt to study cultural evolution solely from the perspective of biology (Dawkins 1976, Wilson, 1998). Some linguists argue for the biological nature of language and study it as an evolving species (Chomsky, 1981;

Bickerton 1990). In economics, Rostchild (1992) published an initial essay on economics as a biological phenomenon. In epistemology, Tolman (1932) wrote an interesting book on the biological behavior of scientific theories and their evolution.

However, despite these clear signs, biology is far from being accepted as the sole reference in the study of human behavior, and the dominant viewpoint among scholars still considers human behavior outside the mandate of natural laws (Nagel, 1986; Searle, 1998).

The way I tried to convey the unitary paradigm

You probably remember that this was the main idea I insisted on conveying throughout your childhood. Subtly, I took every opportunity to engrain these simple facts in your minds: we are animals; we descend from primates; like any living being, our behavior consists of struggling to survive; and this means we must harm other living beings; someone always has to pay for our success, nothing is free in life, etc.

Knowing that the world around you couldn't accept these facts and that you would be compelled to reject them, which would prevent you from understanding human behavior in the future, my intention was to instill a simple scientific fact in your minds from the beginning: we humans are only living beings and nothing more.

And later, when you began to study the nature of matter, I seized the opportunity to emphasize the material nature of the human species. We, humans, are the result of the interaction of millions of material particles, and nothing more.

All of this was not a whim but the consequence of having sufficient scientific knowledge. There are mountains of scientific data from many different disciplines that allow us to conclude, beyond any reasonable doubt, that there is no other reality or phenomenon in us, humans, that is not the result of the properties of matter. Phenomena like spirit, soul, mind, or thought must be treated as material phenomena, as a result of the interaction of material particles.

In other words, human behavior can only be understood through physics, chemistry, and biology. Any attempt to understand it through different philosophies, psychologies, religions, or politics is just useful substitutes for a true scientific understanding (Dennett, 1995; Pinker, 2002).[10]

---

[10] I remember answering your question about what religion was by saying that religious people were those who couldn't understand or couldn't accept that humans are direct descendants of ancient primates, and I also remember how quickly you understood it.

Even though this fact is well established in modern science, as you know, it is rejected or misunderstood by the vast majority of the population. No matter what scientific research can achieve, people are unable to accept that they are simply living beings, in other words, a material phenomenon.

The general belief is that humans cannot be solely treated from a biological perspective because we have two different natures. One nature belongs to the material world, governed by natural laws (physics and biology), and the other belongs to the spiritual world, governed by spiritual laws. Although people admit some influences between these two worlds, they maintain the idea that the two sets of laws governing them are different and unrelated.

Using this dualistic paradigm, people assign genetic phenomena to the material world, while cultural phenomena are considered to belong to the spiritual world. In doing so, they persist in ignoring that all cultural phenomena are the result of the electrochemical activities of neural systems. In other words, they persist in ignoring that culture, soul, spirit, or mind are only the result of the complex activity of the atoms and molecules in our brain.

The systematic and persistent failure of what is called "human behavioral sciences" is due to holding onto this dualistic paradigm. Their belief in spiritual phenomena prevents the full understanding of behavior and, consequently, of these pages as well.

Making an analogy with the past, before the 15th century, astronomers believed that the universe was divided into two main regions: the Earth and celestial bodies. The Earth was governed by laws of corruption, while celestial bodies were governed by laws of purity. However, after the scientific revolution, scientists understood that the entire universe was governed by the same laws. Similarly, today many people think that humans behave under different laws than other living beings due to their spiritual nature, although modern scientific evidence clearly concludes that humans behave exactly under the same laws as the rest of living beings because humans are only living beings.

Main consequence of the unitary paradigm

As you well know, in our universe, no particle can accelerate without the action of an external force (Newton, 1687). No particle can accelerate by itself. This means that all changes in the world are the result of the action of external forces on everything that changes. Consequently, if we assume the unitary paradigm, we must assume that human behavior must be the result of the action of external forces and not of one's own action.

This way of understanding the world is called "determinism" and is the core of modern science. Of course, you are well aware that this is the main obstacle that prevents the idea that human behavior can be completely understood by modern natural sciences from being accepted. The unitary paradigm obliges us to understand our behavior as something completely determined by external forces, and this perspective scares most people.

Although you are perfectly aware of the problem, let me describe what kind of situations the unitary paradigm must overcome to be accepted in the future.

Renowned authors like Berlin (1954, 1958) or Hayek (1944) have argued that this paradigm would lead our democratic societies to become dictatorships where individual and group social freedoms could not exist. According to them, this approach would provide the best justifications for politicians and the military to disregard all laws that protect our personal and social freedoms.

Some philosophers like Heidegger (1927) or Sartre (1947) thought that this new perspective would radically change our sense of freedom and leave our lives without meaning. They believed that in this case, life would be gray, without excitement about the future and the unexpected. There would be no room for creativity or improvisation.

Other individuals like Midgley (1983) or Levinas (1961) argued that the unitary paradigm would devalue and degrade human life, equalizing it with the value of other species. This scenario would open the door to all sorts of manipulation of human beings, just as we are doing with other species, and would lead us to a dehumanized and depersonalized society.

In summary, many people believe that understanding human behavior from the perspective of the natural sciences would mean the end of the human species, a slow but irreversible regression of civilization that would inexorably lead us to self-destruction.

I can assure you that if any of these things were truly possible, I wouldn't mention anything about the unitary paradigm, I would forget it all, and of course, I wouldn't write a book about it.

It's because I am confident that this paradigm will lead us to more cooperative and free societies, where health and well-being will significantly improve, where creativity will expand to astonishing levels, and it will enhance the survival of the human species, that I am writing this book for you and for future readers. Allow me to try to explain why all these fears are unfounded.

Firstly, I would like to mention something that would encourage people to give this new paradigm a chance: the truth has never been a danger to human development. From our history, we have learned that the truth, though unpleasant in most cases, has always brought better results than previous paradigms. Copernicus' theory (1543) yielded better results than the Ptolemaic theory, even though it was difficult to accept. Newton's theory (1687) is the direct cause of our current intense cultural and social development. Pasteur's theory (1861) revolutionized medicine and had a great impact on overall health. Darwin's theory (1859) on evolution was also challenging to accept at the time, but it has led to a greater understanding of nature and had a profound impact on science and culture in general.

Secondly, the scientific foundations of the unitary paradigm are now irrefutable. A wide range of results from many different disciplines, as I outlined before, converge in recognizing humans as a product of natural forces. As research progresses, more and more evidence appears to support this approach's consistency. The denial of this paradigm can only be attributed to scientific ignorance or fear, and not to a lack of sufficient empirical evidence.

Therefore, I believe that any scientist should give the unitary paradigm a chance despite the negative emotions it may initially produce. In fact, most scientists agree with these empirical data and privately believe it is very likely that humans are simply the result of the evolution of life. However, they are very cautious or even reluctant to clearly take this stance and assume all its consequences. Scientists tend to leave open the possibility of some kind of spiritual nature in humans. Today, I realize they do this because of people's fear of anything that could imply an identity between humans and animals. As I mentioned before, people believe that understanding humans on the same level as other species would lead to the destruction of our society. Novels and films periodically explore this theme and force us to imagine a world filled with violence, savagery, and lawlessness, where a few powerful and callous individuals manipulate the masses.

These prejudices primarily stem from a longstanding misunderstanding of Darwin's law. It is not easy to achieve a true and objective understanding of the processes of evolution. After Darwin published his "Origin of Species," where he explained that living beings are subject to the universal law of the struggle for existence, some people saw an opportunity to use this knowledge to their political advantage. The most famous among them was Herbert Spencer (1876). His arguments in favor of social Darwinism were based on the erroneous interpretation of the law of the struggle for life as a justification for discrimination and abuse of power:

"If Darwin's law is true, meaning only the strong survive, there is no point in human societies caring for the sick and disabled." Or, "since progress is a natural outcome of evolution, there is no need for any social intervention or limitation of human activities. Law and authority are fallacious because instead of aiding human progress, they hinder it."

These ideas shared the same misunderstanding: "Current social and political organizations go against Darwin's evolutionary principle. Therefore, it is necessary to correct and guide political action so that society aligns perfectly with Darwin's principle." For example, societies that care for weak individuals go against Darwin's principle, or societies that restrict individual activities also go against Darwin's principle, and so on.

Apparently, for those without a strong background in biology, these ideas, although dreadful, seem to have a scientific basis. What was the mistake of Spencer and his colleagues? Simply forgetting that they, like all living beings, are under the influence of Darwin's law and not above it. No living being has the power to act against Darwin's law, just as no object can defy Newton's law of inertia. It is essential to understand that scientific laws and principles are rules that govern the behavior of things without exception within their scope. Darwin's law is only valid for living beings, which means that all living beings continually behave in accordance with this law. If a single exception is found, Darwin's law will cease to be a law.

The early proponents of these ideas completely misinterpreted the meaning and nature of Darwin's principle. Firstly, because they thought that humans had the power to direct the course of evolution, meaning that humans were above Darwin's law. Secondly, because they did not understand that regardless of the type of society, society is always the result of Darwin's principle, so no political action is needed to adjust societies. In essence, they did not grasp that all societies, from the past to the future, are subject to Darwin's principle.

For example, if a society cares for the sick and the weakest individuals, it means that this behavior is the result of the society's adaptation to its environment. If a society develops many laws and rules to restrict individual actions, it is also the result of the society's adaptation.

In summary, it makes no sense to say, "such a species behaves against Darwin's principle or is not well-adjusted to Darwin's principle, so it needs to be changed to comply with it." Applied to human societies, it makes no sense to say, "such a society behaves against Darwin's law, so it needs to be changed." Not only is it a misunderstanding of Darwin's law, but it is also a social and political danger!

We are not above the laws of nature

It is necessary for you to truly understand what this new paradigm means and what it does not mean. The unitary paradigm states that human behavior is determined by the universal laws of physics, such as the conservation of energy, the increase of entropy, gravity, the laws of electromagnetism, the laws of optics, chemical laws, etc. This means that both current democratic societies and past dictatorships are the result of these laws. It also means that everything beautiful and creative that humans have done is the result of these laws, just as everything terrible and cruel they have done is. It doesn't matter what humans have done or will do; absolutely everything is determined by natural laws.

The same can be said for the behavior of lions, for example. Everything they do, from helping each other care for their young to hunting and eating a buffalo while it's still alive. Kindness and cruelty, creativity and destruction, knowledge and ignorance, everything that happens in the universe is the result of natural laws.

What I'm trying to tell you is that humans do not have the power to rise above natural laws. Our destiny is in their hands. No one can use them at will or for their own benefit. They are supreme forces, completely beyond our reach, that determine how we behave.

As a very simple example, you can consider the force of gravity. None of us can escape it as if we were magicians. Everything we do is under the influence of gravity. We cannot avoid the effects of gravity, although we can fly, swim, or even travel to space like astronauts, where gravity is much weaker. In no case can we eliminate or control gravity. This force will always be beyond the power and will of human beings. Of course, as we learn more about gravity, we can expand the range of things we can do, but we can never have power over it.

The exact same thing happens with all natural laws. We can understand them and use that knowledge to overcome new challenges, but we are always subject to their action, and there is nothing we can do to avoid it. We cannot create or change natural laws. If we were able to do that, we would be gods instead of a living species.

The result of the unitary paradigm

As a result of everything we've discussed here, accepting this new paradigm simply means integrating the human phenomenon into the universal realm of natural phenomena. It's just the final step that human culture must take to achieve a unified materialistic view of the Universe as a whole.

There is only one Universe (reality) which is composed of material things governed by a small set of universal laws under specific circumstances.

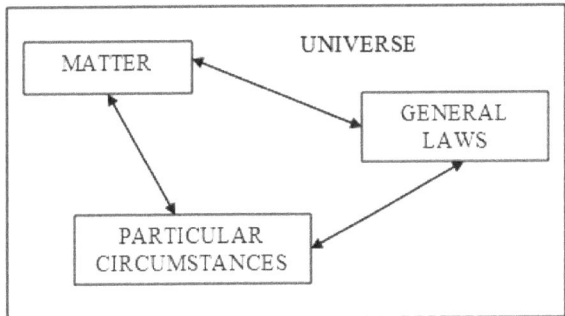

Figure 1. The Universe is composed of matter governed by general laws in particular circumstances.

In other words, we understand human behavior simply as the behavior of matter governed by general laws in particular circumstances. Although you might consider this view highly abstract, it is of utmost importance to keep it in mind throughout your life whenever you attempt to understand a specific issue regarding human behavior.

If you forget that you are dealing with matter governed by universal laws under particular circumstances, you won't be able to reach a true understanding.

When I contemplate death, I try to see myself as a result of the properties of matter, to perceive my experience simply as a physical phenomenon, a material reality like the light that emerges from an incandescent filament when a flow of electrons passes through it. Then I can feel how ephemeral my existence is.

The structure of this book

From the perspective of the unitary paradigm, the study of human behavior should be divided into two main parts. The first part should be devoted to the study of living beings in general. That is, the study of the general laws that govern the behavior of all living beings, with a particular emphasis on animals. Once we have a solid understanding of the general laws that govern the behavior of living beings, we can move on to the second part, which is the study of human behavior.

I would prefer to skip the first part, but unfortunately, there are still no general books on this topic. The study of the behavior of living beings from a physical perspective is just beginning.

This simple two-part plan reveals the main paradigm that underlies all of this work: human behavior is a particular case of the behavior of living beings. From a scientific point of view, this plan is simply evident. However, most people (even most of those who call themselves scientists) reject it due to a lack of extensive scientific training.

This book is written for people who see this plan as self-evident. If the reader has the slightest doubt about it, I encourage them to leave this book and dedicate their time to other matters. If they persist in reading it, they will unnecessarily experience many discomforts and even strong repugnant feelings, wasting their valuable time.

This two-part plan implies that if someone were to write a general book about the behavior of ants, the first part of their book would address the same contents as the first part of this book, suitably adapted to examples of ant behavior. The same would apply to any other species. Conversely, the second part would be very different for each species.

The fact that humans behave under the same laws as other species does not mean that their behavior is the same or similar to that of other species. Each species is unique and usually very different from others, depending on its genetic distance.

In the case of the human species, these differences are maximized due to the emergence of a new biological phenomenon: culture. Although our genetic distance from primates is very short, which means that our physiology is very similar, our behavior is very different. The emergence of language triggered the explosion of a new phenomenon in the history of life that we call "culture" (Dawkins, 1976; Diamond, 1992).

Culture has produced a wide range of behaviors never seen before. The complexity of human behavior has hidden the true nature of humans, making us believe that we were really different from other species. But as I mentioned, we are not different in nature, but we are much more complex and handle much more information than any other species. The second part of this book is dedicated to understanding the main problems of our behavioral complexity.

PART 1: Biophysics

Theoretical bases of behavioral phenomena

"There is nothing more practical than a good theory"
(Kurt Lewin)

Describing behaviors does not mean understanding them. Nothing can be done with thousands of pages describing behavioral facts if one does not have a theoretical foundation to analyze and relate those behaviors.

For many centuries, astronomers collected thousands of data points on the motion of celestial bodies. But despite such a wealth of data, they could not understand the nature of these motions. It was with Newton's gravitational theory that all of these data became understandable; suddenly, many different and seemingly unrelated phenomena appeared coherent and closely related through a simple law.

Exactly the same applies to behavioral phenomena. We need to find a consistent theoretical foundation that makes our observations clear and understandable.

Thermodynamics of life [11]

As modern science evolves, the phenomenon of life is slowly emerging from the darkness of superstitions and misunderstandings. From the early studies of anatomy in the Renaissance (Vesalius, 1543), collections and natural history museums fueled by numerous expeditions around the Earth during the 18th and 19th centuries (Humboldt, 1849), the development of biochemistry shedding light on the details of cellular metabolism (Buchner, 1903), to modern nuclear techniques that allow us to map the molecular structures of genes (Watson & Crick, 1953), living beings are now viewed as material phenomena governed by known natural laws.

Our interest here is to achieve the clearest, simplest, and most general understanding of life that the current state of scientific knowledge can provide. Although specific data about particular species is always welcome, it's difficult to apply them to human cases without an understanding of the basic principles of life. We need the most abstract level possible, the deep logic of life, to be able to apply it to the phenomenon of human behavior.

As you know, physics offers the most basic and abstract level of knowledge currently available, and all natural sciences evolve seeking deep connections with it. Fortunately, in the mid-20th century, a new idea emerged with a unified view of living beings as thermodynamic systems, integrating the entire empirical and disconnected set of known facts. The first to outline such an idea was the eminent physicist Erwin Schrödinger, who had previously developed Functional Quantum Mechanics.

---

[11] The collection of texts in this unit was written in the year 2007 and is part of the document "Scientific foundations of human behavior." This document served as the foundation upon which this book was constructed. (Editor's Note)

Entropy and the second law of thermodynamics

The second law of thermodynamics and the concept of entropy were the result of a brief but exciting period of scientific research initiated by Sadi Carnot in 1824 and culminated by Ludwig Boltzmann in 1872.

The industrial revolution of the 18th century allowed for the transformation of large quantities of stored energy in inert matter (wood, coal, gas, oil, radioactive material...) into useful work. It was based on the development of steam engines, which were a specific type of heat engines. Surprisingly, steam engines were developed without proper scientific backing, relying solely on trial and error. After 1712, Newcomen engines were sold in the UK and some European countries for pumping water from coal mines. Between 1775 and 1800, the firm Boulton & Watt produced 325 rotary steam engines, with 114 of them used in textile industries and 92 in cotton factories. During the first half of the 19th century, hundreds of steam engines were built in Europe to power factories and mines, sparking both a scientific and political, economic, and military race among the brightest minds of European powers.

It wasn't until 1824 when the young French scientist Sadi Carnot published the first treatise on heat engines, laying down the foundation for a future understanding of their nature. Although he based his work on the erroneous concept that heat was a type of massless fluid called "caloric," he hypothesized that the quantity of heat was conserved and that a heat engine produced useful work because caloric circulated from a hot source to a cold one. In addition, his research led to the discovery of an intrinsic loss of efficiency in the conversion of heat into work, unknowingly setting the theoretical groundwork for understanding all forms of energy transformation.

Thanks to the experimental work of J.P. Joule and the theoretical contributions of Lord Kelvin, the German physicist Rudolf Clausius analyzed and clarified the previous research, establishing the fundamental laws of thermodynamics in 1850. Clausius liberated

Carnot's theory from the mistaken concept of caloric, realizing that heat was not a substance but a form of energy transferred between two systems. Moreover, based on Joule's experiments that confirmed the non-conservation of heat, Clausius reformulated Carnot's idea that the quantity of caloric was conserved, stating that what was conserved was energy, not heat.

Die Energie der Welt ist konstant.

The energy of the world is constant, meaning energy is conserved; it is neither created nor destroyed, it simply transfers or transforms from one form to another. This is the popular expression of the first law of thermodynamics. Mathematically, we can formulate this law based on the equation for the change in internal energy in an isolated system. Since work and heat are the only two ways in which energy can be transferred between two systems[12], the change in internal energy is expressed as:

$$\Delta E = Q - W$$

Where Q is the amount of heat transferred to the system from its surroundings, and W is the amount of work done by the system on its surroundings, taking into account the sign convention when defining the thermodynamic system. This principle of conservation of energy seems to promise a symmetrical and stable world, where nothing really changes. Energy, which remains constant, appears to move perpetually, without beginning or end, in any direction, transforming into heat or work interchangeably. However, while experimenting with the ability of heat to produce work, scientists discovered a puzzling and disappointing phenomenon: although work could be completely converted into heat, heat could only be partially converted into work.

---

[12] Excluding the exchange of matter (which was accepted as a form of energy after Einstein).

In this way, Clausius realized that Carnot's research outlined a second law, a fundamental asymmetry in nature that shows the unique direction of all natural processes. Carnot's discovery that there is an intrinsic loss of efficiency in the conversion of heat into work led Clausius to propose in 1865 the term "entropy," and the symbol S, to designate the amount of degraded energy that gradually dissipates through the system's boundary and cannot be used to perform work:

$$dS \ = \frac{\delta Q}{T}$$

In the equation above, we observe Clausius's definition of entropy, which refers to the change in entropy in a system that has reached thermodynamic equilibrium. In this formula, dS is the symbol for the differential change in entropy, δQ is the infinitesimal amount of heat transferred in the process, and T is the absolute temperature at which heat is transferred. This initial definition of entropy as the amount of degraded energy or energy that cannot be used to perform work served Clausius to formulate the popular expression of the second law of thermodynamics:

Die Entropie der Welt strebt einem Maximum zu.

The entropy of the world tends towards its maximum. In any natural process, the entropy of an isolated system always increases. Although the total amount of energy must be conserved, the distribution of this energy changes irreversibly, showing us that the degradation of energy is a requirement of all natural processes. Furthermore, this second law allows us to establish the directionality of all natural processes because the condition that entropy must increase compels nature to always move in this direction. Hot bodies cool down, but cold bodies cannot spontaneously heat up; a bouncing ball eventually comes to a stop, but no stationary ball spontaneously starts bouncing; if we throw a glass onto the ground, it will shatter, but pieces of glass will not spontaneously transform into a glass. Natural energy flow processes cannot go in any direction; they always occur in the direction of increasing entropy, i.e., in the direction of energy degradation.

In this sense, we can observe that the second law inexorably leads all systems to the point where entropy is maximum because it always keeps increasing. When an isolated system reaches maximum entropy, we say it is in a state of thermodynamic equilibrium. In this state of maximum entropy, energy has been degraded almost completely and is uniformly distributed throughout the system, making it unavailable for performing any useful work.

In other words, the farther a system is from its thermodynamic equilibrium, the more capacity it has to increase its entropy and thus has more energy capable of doing work. Conversely, the closer a system is to its thermodynamic equilibrium, the higher its relative entropy and the lower its capacity to perform work because its energy is more degraded and evenly dissipated within the system.

Several years after Clausius's works were published, the Austrian physicist Ludwig Boltzmann introduced a new approach to entropy using arguments from statistical mechanics. Embracing the new atomistic view of matter, he postulated that whenever energy transforms, it shifts from a less probable form to a more probable form.

With his statistical definition, Boltzmann defined the absolute entropy of a system as a function of the number of microstates that could potentially constitute a particular macrostate and demonstrated that this definition was mathematically equivalent to Clausius's thermodynamic concept, thanks to a constant factor known as Boltzmann's constant. In this way, he posited that the amount of entropy in a system is proportional to the logarithm of the number of possible microstates,

$$S \; = \; k \cdot \ln W$$

In Boltzmann's equation, the symbol S represents entropy, k is Boltzmann's constant, and $\ln W$ symbolizes the natural logarithm of the number of possible microstates of the system. The more possible microstates there are for a particular macrostate, the higher the entropy of the observed system. In this sense, entropy indirectly reflects the amount of uncertainty about a system: the greater the number of plausible microstates, the higher the uncertainty about which microstate currently defines the system.

Furthermore, thanks to this statistical definition, we can indirectly relate entropy to the degree of order of particles in a system. A system with higher entropy is closer to thermodynamic equilibrium, and its particles are more disordered, meaning its state is more probable. On the other hand, a system with lower entropy is farther from thermodynamic equilibrium, its particles are more ordered, and its state is less probable. In essence, the second law of thermodynamics could be rewritten as the natural tendency of systems to move from a state of higher order to one of higher disorder or from a less probable state to a more probable one.

In summary, Clausius's definition provides experimental verification of entropy and is solely used to measure its change in an equilibrium system, while Boltzmann's statistical definition extends the concept of entropy to systems not in thermodynamic equilibrium, although under very limited conditions, and provides a new perspective on the nature of the entropy concept.

Unfortunately, the quantitative and rigorous use of the entropy concept cannot be accurately applied to complex biological systems of living beings. Due to this limitation, the use of the entropy concept in the analysis of the behavior of living beings that I will conduct below must be done by extending its rigorous definition to an intuitive approach that allows for qualitative analysis.

Schrödinger's idea about life

The physical nature of life as a phenomenon as a whole ceased to be an enigma since the mid-20th century. Before that, scientists believed that life could not be explained by known physical laws and argued that a new law was needed to account for living beings. They thought that a special type of energy called "vital energy" was responsible for living systems (Bergson, 1907). This special type of energy would be responsible for the apparent self-initiation shown by living beings. However, no empirical evidence of this vital energy was ever found.

In the mid-20th century, a new idea began to replace the old ones. Erwin Schrödinger in 1944 first formulated a coherent explanation of life in terms of known physical laws in his book "What is Life?"

You should have studied this book during your scientific training in biophysics, but if that's not the case, you should proceed to study it now before continuing to read the following pages. The first five chapters of this book are dedicated to introducing the necessary ideas to reach a proper conclusion. The short Chapter Six (Order, Disorder, and Entropy) is devoted to drawing its conclusion, while the final chapter tries to reassure his contemporaries by asserting that, despite his proposal, he did not reject the traditional view that life is governed by unknown laws.[13]

Unfortunately, the fact that Schrödinger was an eminent physicist, a father of modern quantum mechanics alongside Heisenberg, contributed to overshadowing the significance of his brief and singular work in biology, published when he was 57 years old. Once again, in our highly specialized and compartmentalized world, it was unbelievable that an old and eminent physicist could make a new eminent contribution to the field of biology. In other words, being a stranger in that field, Schrödinger's ideas were not met with the interest they deserved from the community of biologists.

---

[13] Science is not an 'angelic' activity but a human activity like any other, subject to the same laws. It is not uncommon in the history of science for eminent authors to make concessions to avoid the strong wrath of their contemporaries.

His idea was simple and clear. Utilizing Boltzmann's discovery (1896) about the mechanical nature of entropy, he considered living beings as systems whose thermodynamic state was extremely far from equilibrium. Since the entropy of living beings increases due to their internal chemical reactions, he concluded that they remain alive by "feeding" on the negative entropy of their surroundings in such a way that the second law of thermodynamics would be satisfied.

As you know, the second principle states that all irreversible processes result in an increase in the entropy of the universe. So, the apparent violation of this principle by living beings results from not considering what happens in their surroundings. The reduction in the entropy of living beings, $\Delta S^l < 0$, is always compensated by a significant increase in the entropy of their surroundings, $\Delta S^s > 0$, resulting in a global increase in the entropy of the universe, ΔS^u, which satisfies

$$- \Delta S^l < \Delta S^s > 0 \qquad (5)$$

and

$$\Delta S^l + \Delta S^s = \Delta S^u > 0 \qquad (5)$$

which means the second law of thermodynamics.

In other words, Schrödinger proposed that living beings maintain their state of low entropy by continuously taking in energy and matter from their environment. This constant flow of energy and matter allows them to sustain their metabolic processes, maintain their organization and complexity, while also increasing the entropy of the universe as a whole. This seemingly simple idea provides us with enough information about the physical nature of living beings to allow for a general analysis of their behavior and leads to some initial conclusions:

1) Living beings and their environment form an inseparable unit, so it is not possible to understand the behavior of a living being without analyzing it within its corresponding environment, as it depends intrinsically on it. Therefore, generalizations about behavior are of very little use. Each living being is and behaves according to its particular environment.

2) Living beings thermodynamically behave as systems that transfer their entropy increases to their environment in order to maintain themselves in a state of nearly stationary relative entropy, and in doing so, they also generate an additional increase in entropy in their environment (5). This formulation will be called 'Schrödinger's theorem,' as he was the first to formulate it in 1944. This principle can be considered the fundamental principle or law of the behavior of living beings.

Living beings and thermodynamic equilibrium

From this perspective, it is clear that living systems are in highly improbable states. Their molecules are arranged in an almost maximum order. If we were to measure the probability of a living being emerging from a random arrangement of its particles, we would arrive at an extremely low value (Kaufmann, 1993; Monod, 1970; Morowitz, 1968). It is almost impossible for a being like us to appear after a random arrangement of our molecules.

You can imagine this probability by considering how rare life in the universe is. This is also why reproduction plays a fundamental role in the cumulative increase in the complexity of living beings (Darwin, 1859; Dawkins, 1976). If every living being had to emerge from inanimate matter, i.e., from the very beginning of life, no complex living being could appear. Reproduction allows each new living being to start from where their parents left off in life. Through very small steps, complexity can accumulate over generations. Although we cannot calculate such a probability, it is evident that it must be extremely low.

Being extremely far from thermodynamic equilibrium means that the entropy of a system is extremely low compared to the maximum entropy it could achieve based on its energy (mass, volume, and temperature), i.e.,

$$S_{rel} = \frac{S_{abs}}{S_{max}} \approx 0 \qquad (6)$$

The relationship between a system's absolute entropy and its maximum entropy gives us the system's relative entropy. Relative entropy is more useful than absolute entropy because living beings and their processes vary widely in size, from processes involving a few molecules to processes involving large societies of individuals. Since absolute entropy is related to the size of a system (an extensive parameter), we cannot use it to study life as a general phenomenon.

On the other hand, relative entropy is independent of the size of a system (an intensive parameter), so we can use it to compare different living beings.

As we will see here, what matters for living beings is not their absolute entropy but how far they are from thermodynamic equilibrium, and relative entropy provides us with such a measure (Prigogine, 1980; Prigogine & Stengers, 1984).

Therefore, we can conclude from deep observation[14] that living beings are systems with extremely low relative entropy. This becomes evident when we observe what happens when a living being ceases to live. In a short period of time, most of its cellular structures disappear (tissues, organs). If we isolate a living being, it will quickly reach its thermodynamic equilibrium. In other words, when a living being dies, its relative entropy increases rapidly until it reaches its maximum value, 1. This experience demonstrates that living beings are far from their thermodynamic equilibrium since thermodynamic equilibrium is defined as the final state a system reaches when isolated.[15]

Es un error común definir la vida como aquello que se encuentra en macroestados muy ordenados. Si bien es cierto que los seres vivos son sistemas muy ordenados, también lo es que los cuerpos inertes pueden estar en macroestados muy ordenados (Ball, 1999). Solamente debéis pensar en un ser vivo que acaba de morir y se ha congelado. Aunque tiene un macroestado muy ordenado, se encuentra en equilibrio termodinámico. Lo que significa que su entropía absoluta habría alcanzado un valor máximo, ya que el sistema (el ser vivo) ha dejado de funcionar y su estructura molecular ha quedado fija y ordenada en el estado sólido del hielo.

It is also an error to define living beings as ordered systems that can grow because crystals are ordered systems that can grow through the process of crystallization (Kauffman, 2000). During crystallization, molecules of a substance arrange themselves repeatedly and regularly

---

[14] The empirical knowledge we have about the complex structure of living beings, from their molecular structures to their ecological interactions.
[15] Isolated simply means that the system does not interact with its surroundings.

in a geometric pattern to form a crystal. However, as you know, crystals are not alive because they are in thermodynamic equilibrium.

Nevertheless, having an extremely low relative entropy is also not sufficient to characterize living beings. There are inert beings that have an extremely low relative entropy, especially stars and galaxies.

Stars have a very low relative entropy not because of their complex structure but because of their extremely high concentration of matter and energy without being isolated. At some point, they also "die," dispersing all their matter and energy, disappearing, and reaching thermodynamic equilibrium to allow new stars to be born.[16]

So, what is the difference between living beings and stars? The difference is the key to understanding the uniqueness of living beings. Living beings exhibit significant fluctuations in their relative entropy, whereas stars do not. The relative entropy of a star always increases,[17]

$$dS_{rel}^{star} > 0 \qquad\qquad (7)$$

even though very slowly, while the relative entropy of a living being increases and decreases, fluctuating over time until it dies. The following figure (Fig. 2) illustrates this concept.

---

[16] This suggests to us that stars and galaxies are also living matter, but their processes are very different from those of living beings on the Earth's surface.

[17] Only when the core of a massive star compresses and heats up before giving rise to a supernova does a decrease in the relative entropy of the core occur. By compressing the core, particles become denser and closer together, reducing the number of possible configurations of the particles, resulting in a decrease in entropy. However, this decrease in the core's entropy is very small compared to the increase in entropy that occurs during the supernova when a massive amount of energy is released. (Editor's Note)

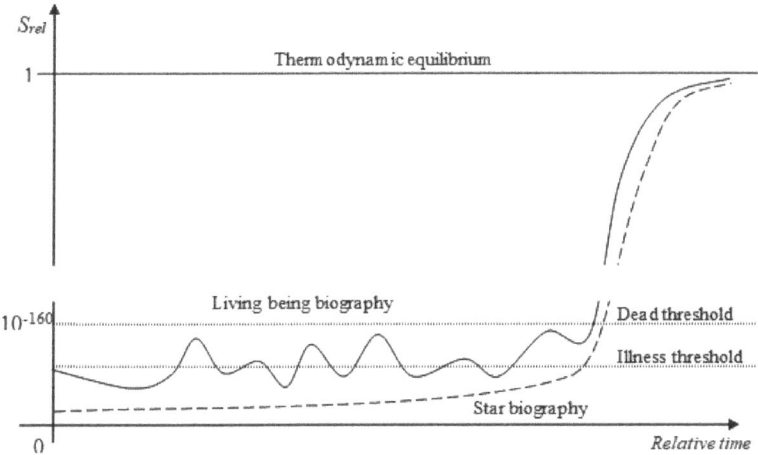

Figure 2. Differences between a star and a living being in their evolution towards thermodynamic equilibrium. The relative entropy of a star always increases, while the relative entropy of a living being fluctuates, showing significant decreases. The term 10^-160 is only used to note that living beings and stars have extremely low relative entropies, but I cannot decipher their actual values.

Stars and living beings stay extremely far from their thermodynamic equilibrium due to different processes. Nuclear reactions in the cores of stars and the force of gravity keep stars stable for billions of years, slowly dissipating a small fraction of their energy into their surroundings, which gradually increases their relative entropy (Adams & Laughlin, 1997). On the other hand, living beings are highly unstable and quickly reach their thermodynamic equilibrium if isolated from their environment. Therefore, living beings depend entirely on their environment to stay away from thermodynamic equilibrium. Living beings use certain interaction processes with their environment to counteract their intrinsic chemical instability, resulting in significant fluctuations in their relative entropy, which spontaneously increases or decreases through a particular class of interaction processes.

Therefore, we can define living beings as open systems that are extremely far from their thermodynamic equilibrium under the influence of certain interaction processes that reduce their relative entropy,

$$S_{rel} > \approx 0 \quad \text{and} \quad \Delta S_{rel}^{interaction} < 0 \qquad (8)$$

Once again, I realize that this conclusion is based on the simple observation of the phenomenon of life. It is easy to see that all living beings need to interact with their environment in a particular way to stay alive (eating, breathing, etc.). It is also evident that when they stop interacting with their environment, they quickly change their state until they reach thermodynamic equilibrium (they die and decompose). So, this is nothing more than a translation into thermodynamic language of what is generally already known about living beings.

Some people have argued against this explanation by saying that certain devices like photovoltaic cells or automatic watches reduce their relative entropy due to specific processes (exposure to light or self-winding) but are not living beings. Unfortunately, these individuals fail to realize that these devices are only constructed by living beings and, therefore, are also related to the phenomenon of life (what Dawkins (1982) called the "extended phenotype"[18]), meaning they couldn't exist without living beings. They also fail to realize that these devices are not far from their thermodynamic equilibrium; they are close to equilibrium, and their processes are governed by linear equations. Although I will delve into this topic further later on, complex living beings act to reduce the relative entropy of the "external" devices they need to survive, such as nests, burrows, tools, etc. Of course, Homo sapiens takes a significant step in this behavior

---

[18] The extended phenotype refers to the idea that the phenotype is not limited to physical characteristics but also includes the impact of that organism on its environment and on the objects it constructs or uses. The tools used by humans can be considered part of their extended phenotype. (Editor's note)

by constructing a large number of "external" devices they need to survive. However, such a phenomenon requires extensive discussion, which I will address later.

Schrödinger Relationship: Negentropic Processes

The key to understanding the phenomenon of living beings is to explain how they can decrease their relative entropy. Relative entropy decreases when the change in entropy of the system is less than the maximum change in entropy of the system.

$$\Delta S_{rel} < 0 \Leftrightarrow \Delta S < \Delta S_{max} \tag{9}$$

The maximum entropy is defined as the entropy that a system reaches at its thermodynamic equilibrium and is directly dependent on the number of particles and the temperature of the system. For living beings, we must consider the number of particles that are chemically stable at the standard temperature when isolated from the environment.

The most massive and energetic systems in thermodynamic equilibrium reach higher entropy values, and vice versa. Therefore, relative entropy can decrease by increasing the mass and energy of the system in such a way that the increase in the maximum entropy of the system would be greater than the increase in the entropy of the system.

This is what happens in living beings thanks to a unique and astonishing phenomenon in the universe. As you well know, the structure of living beings can recover after injury or illness. If you make a cut on a wooden table, for example, the wound will be there forever, but if you cut your arm, it will heal within a few weeks. I like to call this special property "wound recovery." A second special characteristic of living beings is that they can grow even when far from their thermodynamic equilibrium.

However, both properties are nothing more than the manifestation of a single phenomenon: living beings are capable of reproduction, that is, of making copies of themselves. Wound recovery and organism growth are achieved through cell reproduction. The cellular structure is the common mechanism of all living beings[19] and is an astonishing reproductive machine.

These processes of cell reproduction are primarily responsible for the decrease in the relative entropy of living beings since, in wound recovery, they involve the creation of new cells with a more ordered structure and organization, leading to a decrease in the absolute entropy of the system. Additionally, cell reproduction can also contribute to the decrease in relative entropy by increasing its maximum entropy since it involves the creation of new cells that can contribute to increasing the mass and number of particles in the system, thereby increasing the maximum entropy that the system could reach at its thermodynamic equilibrium.

As we can observe in equation (6), there are two ways in which a system can decrease its relative entropy: by decreasing its absolute entropy while keeping its energy constant (wound recovery processes) or by increasing its maximum entropy (growth processes). Thermodynamically, this means that living beings are not only far from their thermodynamic equilibrium but are also capable of achieving significant decreases in their relative entropy.

In this sense, we see how living beings are compelled to act against the general trend of moving toward thermodynamic equilibrium. In fact, they not only resist this trend, like stars, but also move against it, further distancing themselves from their thermodynamic equilibrium by decreasing their entropy.

$$dS < 0 \qquad\qquad (10)$$

Of course, this situation is not continuous and permanent. It belongs to a strong entropy fluctuation in which the total change in relative entropy can be considered negligible for a long period, maintaining the relative entropy extremely low.

---

[19] Viruses do not have a cellular mechanism; they cannot survive without the cells of other living organisms. In this sense, we say they are cell-dependent.

$$\Delta S_{rel} \simeq 0 \text{ and } S_{rel} > \approx 0 \qquad (11)$$

However, eventually, the general tendency towards thermodynamic equilibrium prevails, and all living beings, like stars, quickly move towards reaching their thermodynamic equilibrium.

$$S_{rel} = 1 \qquad (12)$$

In summary, living beings are characterized by meeting conditions (11) that can only be achieved if something can force a decrease in their entropy (10).

Since ancient times, philosophers and scientists have seen living beings as possessing a special nature, different from that of inanimate bodies. Although they couldn't formulate it, they were somehow aware that property (10) went against natural tendencies and, therefore, was a kind of miracle.

Indeed, you know that the second law of thermodynamics states that any process in nature[20] must lead to an increase in the entropy of the universe. This principle defines the general tendency of things to reach thermodynamic equilibrium and even the irreversibility of time (the arrow of time). Considering this law, ancient scientists saw living beings as systems that apparently violated the second law in some way, so they assumed that there must be a third unknown law to account for it.

Schrödinger's idea was that this third law was not really necessary. Instead of introducing a new law, he introduced design as the key to understanding this puzzle. In particular, he formulated that the second law would not be violated if some interaction processes could transfer entropy from living beings to their environment, through their particular design.

---

[20] Of course, processes carried out in isolated systems at their thermodynamic equilibrium do not increase the entropy of the systems (they are called "reversible processes"). Although they are useful for theoretical purposes, in reality, they do not exist in nature.

Stars and living beings also differ in how they are extremely far from their thermodynamic equilibrium. Stars are very far from their thermodynamic equilibrium due to their enormous amount of highly concentrated mass and energy, while living beings are very far from their thermodynamic equilibrium due to their immense complexity.

Considering that the entropy of the universe always increases because any process must produce entropy, the entropy of a living being could decrease if some processes were able to transfer enough entropy from the living system to its environment. In other words, this would be possible if the increase in entropy of the universe is supported by the environment of the living being rather than the living being itself. Behind Schrödinger's idea was the consideration that the second principle only imposed an increase in entropy for isolated systems, while living beings were clearly open systems in which mass and energy were continuously exchanged with their environment.

Taking dS_i^l as the entropy produced by the internal processes of the living system l, taking dS_x^l as the entropy produced by the exchange process x on the living system l, taking dS_x^s as the entropy produced by the exchange process x on the environment of the living system s, and taking dS_x^u as the entropy produced in the universe (living being plus its environment), the second law of thermodynamics states that:

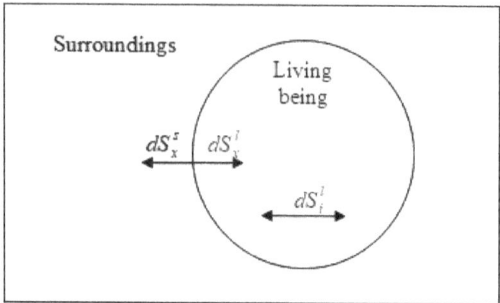

Universe

Surroundings

Living being

dS_x^s | dS_x^l

dS_i^l

Figure 3. Diagram of the thermodynamic universe formed by a living being and its environment in which we can observe the entropy generated in the interaction processes with the environment, as well as the entropy generated by the internal processes of the living being.

Consider the following simple case of a living system l with many internal processes, its environment s, and an interaction process x between the living system and its environment.

The second law states that both the internal processes within the living system and the interaction process must produce entropy,

$$dS_i^l > 0 \text{ and } dS_x = dS_x^l + dS_x^s > 0 \quad (13)$$

On the other hand, we have seen that a living being can only exist if its entropy decreases (10), that is,

$$dS^l = dS_i^l + dS_x^l < 0 \quad (14)$$

Now we need to find a class of interaction processes x that can satisfy both the second law (13) and the existence of living beings (14). From (14), we obtain

$$- dS_x^l > dS_i^l \qquad (15)$$

which means that these interaction processes x must be capable of reducing the entropy of the living system by an amount greater than the entropy produced by the internal processes of the living system. To do that without violating the second law, the interaction processes x must also satisfy

$$dS_x^s > - dS_x^l \qquad (16)$$

translate: which is derived from (13). This means that these interaction processes x must be capable of increasing the entropy of the environment by an amount greater than the entropy they reduce in the living system. You can note that it's not sufficient for the interaction processes to balance the entropy they reduce in the living system. They must produce more entropy in the environment than the entropy they reduce in the living system. Combining (15) and (16), such interaction processes x are characterized by

$$dS_x^s > - dS_x^l > dS_i^l \qquad (17)$$

translate: In other words, the entropy produced by these interaction processes in the environment must be greater than the entropy they reduce in living organisms, which in turn must be greater than the entropy produced in living organisms by internal processes. The type of interaction processes that satisfy (17) does not violate the second law and allows the existence of living beings because it is capable of decreasing the entropy produced in living organisms and, at the same time, increasing the entropy of the Universe.

To understand the nature of the processes that satisfy (17), consider the total entropy produced by such a process and the internal processes of living beings, which is given by

$$dS^U = dS^s + dS^l > 0 \qquad (18)$$

$$dS^U = dS^s_x + dS^l_x + dS^l_i > 0 \qquad (18)$$

but, due to (17),

$$dS^l = dS^l_x + dS^l_i < 0 \qquad (19)$$

then we get

$$dS^s_x > dS^U \qquad (20)$$

In other words, the entropy produced in the environment by the interaction processes x that satisfy (17) must be greater than the sum of the total entropy produced by that process and the internal processes of the living being. It seems clear that the "cost" of the existence of living beings is an additional burden of entropy from their environment (20), which they have to pay (increase their entropy) to preserve the second law and reduce the entropy of the living being.

So, the second law does not prevent the existence of living beings, as ancient scientists believed, but it imposes a strict condition: living beings can exist only if there are some interaction processes x whose particular design allows them to increase the entropy of their environment to a greater extent than the entropy they are capable of reducing in themselves.

The key to understanding living beings is to focus on their special design rather than looking for special types of energy. The occurrence of processes that satisfy (17) depends solely on the particular design of living beings. In this sense, a living being can be defined as the material system whose design allows for interaction processes that satisfy Schrödinger's theorem (17).

Example: aerobic respiration

The simplest way for interaction processes to satisfy Schrödinger's theorem is by coupling two simpler processes. One of them is responsible for a decrease in entropy in the living being, and the other is responsible for an increase in entropy in the environment, where their combined action satisfies (17).

In fact, such simple processes do not exist in nature. Even the simplest living being has an extremely complex design, and its processes are also extremely complex. However, for theoretical purposes, we can reduce complex interaction processes to the combined action of two or more simple processes.

An example of this is the process of respiration in animals, which basically involves the transformation of oxygen and glucose from the environment into CO2, water, and heat, while some molecules of ATP are built from molecules of ADP and phosphate inside the animals. Although it is a very complex process, its simplified form can be written as:

$$C_6H_{12}O_6 + 6O_6 + 36Pi + 36ADP \rightarrow 6CO_2 + 6H_2O + 36ATP + heat$$

This is an interaction process because it simultaneously affects the living being and its environment.

$$\Delta S^{U} = \Delta S_{x} = \Delta S_{x}^{i} + \Delta S_{x}^{j} = 0.79 \, \text{kcal/K} > 0$$

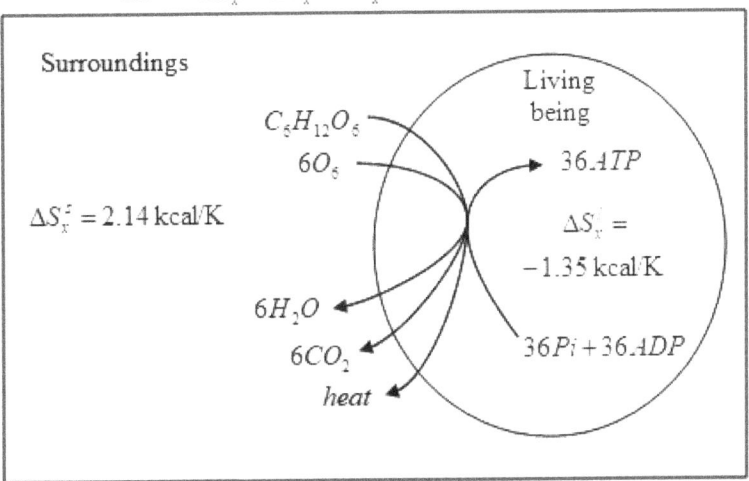

Figure 4. The simplified respiratory process is an interaction process x that satisfies Schrödinger's theorem. Glucose and oxygen in the environment of aerobic living beings compel aerobic living beings to capture and utilize them to synthesize ATP from ADP and Pi, returning water, CO2, and heat to their environment.

For theoretical purposes, we can separate it into two very simple processes, one that affects only the living being and one that affects only its environment.

ATP synthesis in living organisms:

$$36Pi + 36ADP \rightarrow 36ATP \qquad (21)$$

Oxidation of glucose in the environment:

$$C_6H_{12}O_6 + 6O_6 \rightarrow 6CO_2 + 6H_2O + heat \qquad (22)$$

Under standard conditions[21] and normal concentrations in cells, the synthesis of ATP has a free energy[22] gain of 12 kcal/mol. Since one mole of glucose can be used to synthesize 36 moles of ATP, we need to modify the total free energy gain in the synthesis of ATP for one mole of glucose, which means there is a total free energy gain of (36 moles x 12 kcal/mol) 432 kcal for one mole of glucose. This represents a decrease in the entropy of the living system of -1.35 kcal/K per mole of glucose. Of course, this process would violate the second principle and cannot occur in isolation.

For the oxidation of glucose, there is a decrease in free energy of -686 kcal/mol in the environment. This results in an increase in entropy in the environment of 2.14 kcal/molK and does not violate the second principle.

If both processes, (21) and (22), are coupled, as is the case in aerobic animals, it is easy to see that the increase in entropy in the environment more than compensates for the decrease in entropy in living organisms. The total change in entropy of both coupled processes is 0.79 kcal/K per mole of glucose, which complies with the second law of thermodynamics and also satisfies the left side of Schrödinger's theorem (17).

Of course, we cannot verify the right side of (17) because we do not know how much entropy has been produced in living organisms due to other processes. But we could view the process of respiration as one that can contribute to keeping living organisms alive because it is capable of decreasing their entropy.

---

[21] Temperature: 298 Kelvin, pressure: 1 atm and pH: 7,0.
[22] Instead of measuring entropy, chemists prefer to measure the free energy or Gibbs energy (1902), which is related to entropy by $\Delta S=1/T(H-\Delta G)$, where H is the heat absorbed by the system, and G is the free energy of the system.

Overcompensation processes instead of dissipative processes

I'm sure you've read about the thermodynamics of living beings and dissipative processes (Prigogine & Nicolis, 1977; Prigogine 1980; Prigogine & Stengers, 1984). You might be feeling a bit puzzled because my presentation is somewhat different from the current ones. You are correct, and let me explain the nature of this difference.

The disagreement lies in (16), because current theorists write

$$dS^s_x = - dS^l_x \qquad (23)$$

where I write

$$dS^s_x > - dS^l_x \qquad (16)$$

Current theorists like (23) because it allows them to talk about the transport of entropy, $\left| dS^l_x \right|$, from the living being to its environment. However, by doing so, the second principle is not fulfilled because such an interaction process does not increase the entropy of the universe, $dS_x = dS^l_x + dS^s_x = 0$, which is not in thermodynamic equilibrium. A process that involves the transport of a certain amount of entropy keeps the entropy of the universe constant and, therefore, violates the second principle. Perhaps this strategy could simplify some mathematical calculations, but it does not align with the facts and can lead to erroneous concepts.

The second principle states that all processes must produce entropy in a universe far from its equilibrium, and as a result, (23) is an error.

Current theorists call interaction processes that satisfy (23) "dissipative processes" because they can be seen as processes (dissipators) of entropy transport. However, interaction processes that satisfy (16) cannot be seen as simple transport processes because the entropy at the destination is greater than the entropy at the source. So, rather than entropy transport, which sounds a bit off because entropy is a non-conservative quantity, I prefer to call them "overcompensation processes," which better expresses what they do. They overcompensate for the entropy they reduce in living beings with a greater increase in entropy in the environment (16).

In summary, Schrödinger's idea is that living beings stay alive thanks to their design, which allows them to engage in interaction processes with their environment in which they increase the entropy of the environment to a greater extent than the entropy they reduce in themselves. This is the main clue to understanding the behavior of living beings.

To the question: What do living beings do? Our answer must be: they constantly strive to increase the entropy of their environment more than the entropy they reduce in themselves.

Life is a matter of design

Now, we can recap our initial question: What is life? We have seen that living beings are material systems whose particular arrangement of particles allows interaction processes that satisfy Schrödinger's theorem (17).

This idea leads us to understand that life is simply a matter of design or arrangement of particles. Specific designs, under the influence of external forces, behave like living systems. There are no special, new, or particular forces involved in the phenomenon of living beings. It's simply natural forces acting on particular arrangements of matter.

Let me illustrate this idea with a very simple example.

Consider the design differences between a wheel and a box. Imagine that their mass and particles are the same, but they are arranged differently. On a slightly inclined plane, the box remains in place while the wheel can roll along it, under the force of gravity. In other words, the behavior of things under natural laws depends on their design, on how their particles are arranged.

The particular behavior of living beings is not the result of hidden entities but of the design they have. While wheels have a design that compels them to roll on a slightly inclined plane under the force of gravity, living beings have a design that compels them to do what they do under the action of all sorts of natural forces.

Behavior is a matter of design.

Relative entropy gradient as a general force of life

In Schrödinger's theorem, it doesn't mention how the transfer of entropy occurs. In fact, one could still think that some internal energy (vital energy) is responsible for such transfer, maintaining the idea of some kind of vital will. In other words, living beings don't seem to be moved by external forces, as mechanical laws demand, but by their own initiative.

It appears that they are responsible for transferring their entropy to their surroundings. When we observe a cow grazing, it seems like the increase in entropy in the grass is caused by the cow's behavior. However, this is just an appearance. If we think about it carefully, we will see that the primary cause of the existence of living beings is the Sun.

Schrödinger's theorem states that the environment must take on the entropy produced by living beings, i.e., it must compensate for the decrease in entropy in living beings with a sufficient increase in its own entropy to maintain a positive balance in entropy production throughout the universe.

This is what the Sun does spontaneously! The Sun is moving spontaneously toward its thermodynamic equilibrium, meaning it is continuously increasing its entropy in large quantities. No living being forces the Sun to increase its entropy. On the contrary, it is the Sun that forces living beings to exist and behave.

Considering $\phi^{sun} = \frac{dS^{sun}}{dt}$ as the rate of entropy production in the Sun and $\phi^{orbit} = \frac{dS^{orbit}}{dt}$ as the rate of entropy production in the immediate surroundings of the Sun (which can be defined as the space that includes all particles orbiting around the Sun), we can define a generalized force as the difference in entropy production between the Sun and its surroundings.

It is clear that

$$\Theta = \Delta\phi^{Orbit-Sun} = \phi^{Orbit} - \phi^{Sun} <<< 0, \qquad (24)$$

because the matter on planets and satellites is mostly in its thermodynamic equilibrium, where entropy production is practically negligible. The entropy production of a system depends directly on how far it is from its thermodynamic equilibrium.

Because the entropy production in the Sun is extremely high compared to the entropy production in its surroundings, a potential entropy production force is established. The tendency of entropy production to balance between the Sun and its surroundings is the origin of this potential force.[23]

Therefore, the Sun is constantly forcing the systems in its surroundings to produce entropy or, in other words, to move away from their thermodynamic equilibrium. The extent of this movement depends on the exchange processes of each system. In most cases, where the design of the systems (the arrangement of particles) does not allow for dissipative processes, there is a brief departure from thermodynamic equilibrium that spontaneously restores. These can be considered as simple disturbances of equilibrium. In these cases, entropy production is very low. In a few other cases, the arrangement of particles in the systems allows for dissipative processes, and these systems are forced to move away from their thermodynamic equilibrium due to the special ability of dissipative processes to transfer the produced entropy to their surroundings. As these systems move away from their thermodynamic equilibrium, their entropy production increases, and they do not spontaneously return to their thermodynamic equilibrium because their dissipative processes are continuously active under the influence of the Sun.

---

[23] While entropy is not a conservative parameter, many authors agree that it can be treated as the subject of exchange processes with the necessary precautions. I believe that the most important thing is not to forget that entropy is not conservative.

The systems compelled by the high potential for entropy production generated by the Sun are living beings. Because dissipative processes can be simultaneously coupled (parallel coupling) and can be linked in time (series coupling), the persistent potential entropy production force of the Sun has produced a highly complex living system over millions of years of evolution, where the transfer of entropy to the Sun is accomplished through a very complex and sophisticated chain of linked dissipative processes involving a large number of living beings.

In other words, while the direct relationship between the Sun and plants is clear, the transfer of entropy to the Sun by birds, for example, is less evident.

In summary, what mechanics establishes, that an external force is always necessary to produce a change in any system, also occurs in living systems. Living systems (systems whose design or arrangement of particles allows for dissipative processes) are compelled to change (behave) whenever there is a potential for entropy production between them and their environment.

When this potential force takes negative values,

$$\Theta^l = \Delta\phi = \phi^l - \phi^s < 0 \qquad (25)$$

living beings are forced to reduce their entropy (they are compelled to live). That is, when the production of entropy in their environment is greater than the production of entropy in living beings, there is an opportunity to increase their production of entropy, to grow, to repair damage, to learn, to reproduce, etc.

When this potential force takes positive values,

$$\Theta^l = \Delta\phi = \phi^l - \phi^s > 0 \qquad (26)$$

living beings are forced to move towards their thermodynamic equilibrium (they are compelled to die). Therefore, it seems clear that the phenomenon of life is a matter of the particular design of particles (dissipative processes) under the influence of a generalized external force (entropy production potential). It's not the cow that forces an increase in the entropy of the grass by eating it, but the entropy production of the Sun that forces the grass to produce entropy, which in turn forces the cow to produce entropy when eating the grass. In other words, in the presence of grass, the cow cannot avoid eating it.

To analogously visualize that there is no mystery in living systems, you could imagine a waterwheel turning in a river. This phenomenon has two causes. First, the particular design of the waterwheel that allows it to rotate under the action of water currents. Second, the existence of a water current falling onto it. That is, it is the result of a particular design that works under an external force, and nothing more.

I have tried to intuitively show you that a living being is the same: a particular design that functions under an external force.

This matter is of utmost importance because the first difficulty in understanding human behavior is to comprehend that there is no room for freedom and will. Although subjectively, we experience freedom and will, this is just an illusion. From an objective standpoint, all behavior of living beings is the result of the action of external forces on them, just as it is with anything else. Our deep conviction that humans make decisions, meaning that they are not the direct result of external forces but the result of our free will, needs to be overcome.

Behavior of living beings

Life is intimately linked to entropy. The episodes of life are written in the language of entropy. This means that everything we can say about living beings, we should be able to translate into terms of entropy. Therefore, to understand living phenomena, we need to be sensitive to all the ways in which entropy manifests itself to our senses. Fortunately, as we will see later, our senses are specially designed to be sensitive to the forms of entropy, so all we need to do is train them as much as we can.

As I mentioned before, living beings are pieces of extremely and surprisingly organized matter. Their complex design is far from being copied by our engineers. Their shapes are science fiction to our eyes. The immense variety of their forms and solutions seems endless. In terms of entropy, living beings are pieces of matter with extremely low relative entropy.

We have seen that there are two ways in which a material body can be in a state of extremely low relative entropy. One way is to have an extremely high concentration of energy with very few different particles, like stars and galaxies. Their very low relative entropy comes from the large difference in energy density between their interior and exterior. The other way is to have a very large number of different particles (molecules) arranged in a highly ordered manner. Their extremely low relative entropy comes from the high degree of order among the very diverse particles inside them and the less diverse particles outside them.

Living beings belong to the second case. They are material bodies formed by an enormous variety of molecules that are arranged in a highly ordered manner. The probability of these configurations is close to zero. But being extremely ordered is not the most important characteristic of living beings.

Surprisingly, despite living in a world where particles tend to randomize, producing an increasingly disorderly universe, those pieces of matter we call living beings maintain their highly ordered state for a significant amount of time.

The behavior of living beings is described by the set of processes that satisfy Schrödinger's theorem (17). Understanding the behavior of a particular living being involves understanding how each of its specific interaction processes satisfies the theorem. The unit of analysis for behavior corresponds to identifying a specific interaction process that satisfies Schrödinger's theorem, i.e., identifying how the relative entropy of a living being decreases through an increase in entropy in its surroundings. The previous example of aerobic respiration illustrates this. Therefore, respiration can be considered as a unit of behavior.

This is the most important thing we can say about the behavior of living beings, and it is of paramount importance that you can understand it deeply. Describing behavioral facts observed in living beings is of no value without understanding how Schrödinger's theorem is fulfilled.

You should consider the following rule as the universal rule for understanding the behavior of any living being:

Identify:
1) a decrease in relative entropy in a living being.
2) a simultaneous increase in relative entropy in its surroundings.
3) the interaction process responsible for 1) and 2).

Whenever you can satisfy this rule, you will achieve an understanding of a particular behavior.

Qualitative versus quantitative analysis

Unfortunately, the thermodynamics of systems far from equilibrium is not yet well-developed. Currently, we are far from being able to perform quantitative thermodynamic analyses of the behavior of living beings. Measuring the change in relative entropy for living beings and their surroundings remains nothing more than a fantasy.

However, instead of avoiding behavioral analysis due to a lack of measurement capabilities, we can conduct qualitative analyses. Of course, the reliability of qualitative analysis can be very low, and many errors can be made. Therefore, we must take the utmost precautions and never forget that we can fail in our conclusions. Only repetitive training in similar situations and the accumulation of empirical experience can improve our reliability.

In summary, we can only perform behavioral analysis qualitatively, which means our reliability is low. Nevertheless, continuous practice and empirical confirmation can improve our reliability and make our analyses better and more useful.

However, never forget that this qualitative analysis, guided by Schrödinger's theorem, is the best that can be done. Your health depends directly on the analyses you can make of your own behavior and the behavior of the people you interact with. Despite its shortcomings, Schrödinger's theorem points us in the right direction to achieve a true understanding of behavior.

From the perspective of behavioral facts, nothing can be added to Schrödinger's theorem. Our factual knowledge is summarized in this theorem. However, its abstract formulation makes it difficult to recognize it in the wide variety of behavioral phenomena. Now, our challenge is to develop knowledge and skills to apply it to various behavioral situations, and there is much work we can do in this regard. The rest of the first part of this book is dedicated to developing skills to apply Schrödinger's theorem to the analysis of the behavior of living beings. The second part is dedicated to developing skills to apply it to the analysis of human behavior.

Life is a problem of order

The first improvement to apply Schrödinger's theorem is to consider that entropy is also a measure of disorder. Entropy measures the probability of a macrostate of a given system, and this probability is proportional to the number of different particle arrangements that result in the same macrostate (Boltzmann, 1896). When humans have the opportunity to observe the arrangement of particles in the system, we all agree to classify arrangements with low entropy as "ordered" and those with high entropy as "disordered." This fact can be seen as a gift to enhance our behavioral analysis since we are very sensitive to perceiving order and disorder.

In other words, instead of trying to perceive changes in entropy in behavior, we can focus our attention on perceiving changes in order, which is easier for us. When you were children, you couldn't perceive an increase in entropy when you were with me, but you could clearly perceive an increase in disorder in my system, and that's why you could conclude that there was an increase in entropy in my system. Perceiving disorder is how we perceive entropy.

However, we must take an important precaution when observing changes in disorder. As we have seen before, what's important when studying the phenomenon of living beings is not entropy but relative entropy. This means that many orderly things are not alive. For example, a snowflake is highly ordered but not alive. Even if you could see how it forms, you would observe a decrease in entropy from a water droplet to a snowflake. But you should pay attention that its relative entropy does not decrease! Why? Because there has been a decrease in temperature, meaning the system has lost energy (size). The number of available arrangements has decreased due to the reduction in temperature. In other words, its relative entropy remains the same and equals one because it stays in its thermodynamic equilibrium, even though it is highly ordered.

Considering that entropy is also a measure of a system's disorder, relative entropy is a measure of a system's relative disorder. This allows us to conclude that living systems are extremely ordered, which is an observable fact. Although we cannot calculate the entropy of a living being, we can infer simply by observation that all living beings are extremely ordered, and their order disappears when they die. In other words, we can associate life with high relative order. It is important not to confuse order with relative order. Many inert objects have order (absolute order) but do not have relative order because they are in their thermodynamic equilibrium, as we have seen in the formation of snowflakes.

The existence of great order in living beings allows us to construct a complex classification of millions of different species, distinguish each individual within a community, draw detailed anatomy of each species, etc. (Mayr, 1982).

It is also a fact that the loss of life, death, correlates with a clear degradation of the previous order. When a living being dies, there is a significant increase in entropy until it reaches its thermodynamic equilibrium (Morowitz, 1968; Prigogine, 1977). For living beings, entropy signifies death.

Darwin's principle satisfies Schrödinger's theorem

You know that Darwin's principle of the struggle for existence (1859) is the foundation of modern biology and is the best approximation we have for understanding the behavior of living beings. In summary, this principle states that every living being must compete against other living beings to obtain the necessary resources for survival and reproduction, resulting in the selection of species, which, in turn, leads to the evolution of species over successive generations due to variability in their reproduction.

Now, it is important to realize that the principle of the struggle for existence establishes the same concept as Schrödinger's theorem. To see this, we first need to discuss what the best places in the environment of living beings are where entropy can be dissipated, in other words, which systems allow entropy to increase more easily.

There are two main types of things in the environment of living beings: 1) inert bodies and 2) other living beings. With few exceptions, all inert bodies are in thermodynamic equilibrium, meaning they are filled with entropy or store the maximum entropy they can. Therefore, their entropy can only increase by adding energy. For example, the entropy of a rock can increase if it is broken into many pieces.

From Darwin's principle, we see the behavior of living beings as a continuous struggle for resources against other living beings. These other living beings are of their own species (conspecifics) and of other species. Modern community ecology is dedicated to studying the wide variety of living beings that engage in this struggle. Although the study of their behavior is well-established in modern biology, understanding Darwin's principle from a thermodynamic perspective will greatly help us extend this study to human behavior. Mainly because the behavior of our species involves culture, and the best way to understand culture is through the concepts of information theory, which we can introduce through thermodynamics, although we will see this later.

As you may have already guessed, the best places in the environment to dissipate entropy are precisely the other living beings because they are far from thermodynamic equilibrium and can easily store large amounts of entropy. On the contrary, it is very difficult to increase entropy in systems that have reached their maximum entropy. In fact, without increasing their internal energy, they cannot increase their entropy. For example, a rock on a mountain is in its thermodynamic equilibrium. If we wanted to increase its entropy, we could break it into pieces, but the only way to do so would be to increase its energy by performing a significant amount of work on it. However, the increase obtained would be very low.

In contrast, with relatively little effort, we could produce a significant increase in entropy in the living beings in our environment. For example, with a simple match, we could completely burn down a forest. Since a forest is full of living beings, and many of them could not escape the flames, using a match would result in a large number of dead organisms. In this case, we would produce a significant increase in entropy in our environment because the living beings in the forest are far from thermodynamic equilibrium, and they would reach it shortly after their death.

In other words, if all systems in the environment of a living being are in their thermodynamic equilibrium, meaning at their maximum entropy, they will not be able to absorb the entropy produced by the living system. Consequently, the living system will not be able to dissipate any amount of entropy, which will lead to its progression towards illness (inability to dissipate entropy) and ultimately towards death (thermodynamic equilibrium). Therefore, it is essential that there is some system in the living being's environment that is far from its thermodynamic equilibrium so that it can acquire the entropy that the living being needs to dissipate. The farther a system in the environment is from thermodynamic equilibrium, the more likely it is to acquire the entropy that the living being needs to dissipate. Conversely, systems close to equilibrium will not be able to absorb the entropy from the living being.

This means that living beings require highly ordered environments to be able to dissipate their entropy (Prigogine & Nicolis, 1977). However, since living beings are the most ordered systems in their environment, they also constitute the potential reservoirs of entropy for other surrounding living beings. Hence, there is a struggle among living beings to dissipate their entropy, and those that fail to do so ultimately perish in the struggle for existence (Atkins, 2007; Kaufmann, 1993).

Furthermore, as a consequence of this struggle, living beings must dissipate not only the entropy they produce themselves but also the entropy they receive from other living beings in their environment. Therefore, the death of a living being may result not only from its inability to dissipate the entropy it produces but also from the entropy it receives from other living beings in its environment. Modern medicine has established that communicable diseases are caused by attacks from microorganisms such as viruses or bacteria, which feed on the resources of the sick individual. From the perspective presented here, these attacks can be seen as interactions in which microorganisms feed on the negative entropy or free energy of the sick individual, resulting in an increase in their relative entropy.

When Darwin proposed the term 'struggle for existence' to describe a fundamental characteristic of life, he was, without being aware of it, referring to Schrödinger's theorem that characterizes living beings. Indeed, to struggle means to try to produce entropy in the adversary. Schrödinger's theorem can be viewed as the struggle between the living system and its environment to rid itself of entropy. The living system can only maintain itself as such if it is capable of removing negative entropy or free energy from its environment. The environment, in turn, resists receiving entropy from the living system, limiting the possible ways in which it can be brought towards thermodynamic equilibrium.

"The general struggle for the existence of living beings is not a struggle for basic elements (the basic elements of all organisms exist abundantly in the air, water, and soil), nor is it a struggle for the energy contained in the form of heat, unfortunately unusable, in every body, but a struggle for entropy, which is only available in the transmission of energy from the hot sun to the cold earth. To make the most of this transmission, plants open the vast surface of their leaves ..., to carry out chemical synthesis... The products of this chemical process constitute the object of the struggle of the animal world." (Boltzmann, 1896)

As we have seen before, every living being is a system far from thermodynamic equilibrium, subjected to strong internal and external pressures to restore equilibrium, meaning disorder and death. However, it faces these pressures by having mechanisms of interaction with the environment to maintain its state of extremely low relative entropy. Now, can any design work? Obviously not, as the environment limits the designs that can be efficient and useful. For each particular environment, only a small set of designs are valid for successfully performing these interactions. Thus, any design that cannot correctly dissipate entropy in its environment will not thrive, and therefore, it will not reproduce. Only designs that can successfully carry out these interactions in their particular environment will thrive.

This leads to a natural selection of the various replicas that each organism generates. Replicas whose modifications or variations are not suitable for successfully performing the dissipative function will not replicate, and therefore, their characteristics will tend to disappear in subsequent generations. Conversely, replicas that perform it more successfully and efficiently will tend to produce a greater number of new copies, which means their characteristics will tend to dominate in subsequent generations.

Thus, the struggle for existence arises from the degree of efficiency with which organisms can perform the interaction with the environment that satisfies Schrödinger's theorem. The more efficient a particular organism is in dissipating entropy to the environment, the greater its chances of propagating its characteristics to future generations. In reality, natural selection only selects those organisms that can achieve a greater number of replicas in the next generation, but for this, it is necessary for these organisms to be efficient. Efficiency and replication are, separately, necessary but not sufficient for natural selection. Evolution requires the combination of both conditions simultaneously.

However, selection can be more or less intense, more or less strict depending on the pressure to which different organisms are subjected. A population living in a very resource-rich environment will be subject to less harsh and strict selection than a population living in a very poor environment. Due to the nature of Schrödinger's theorem, selection tends to become increasingly strict and intense over time, which is why evolution tends to generate increasingly complex designs.

Indeed, if a population lives in a very rich environment, the selective pressure will initially be lower, but over time, that population will deplete the resources (order) of its environment, leading to an increase in selective pressure. Therefore, selective pressure tends to increase over time, reducing the range of selected characteristics.

Furthermore, increased selective pressure requires more elaborate and efficient designs. The need to access new forms of resources, due to increased selective pressure, necessitates the design of more sophisticated mechanisms. When selective pressure in the marine habitat challenges the existence of a particular species, mechanisms that allow this species to access the terrestrial habitat are selected, meaning those that enable it to become amphibious. When coal extraction was no longer sufficient for industrial maintenance, increased selective pressure generated by industrial competition led to the design of more complex engines based on the combustion of petroleum derivatives. When social conflict in the slavery-based state became unsustainable, selective pressure generated new state forms based on solidarity and equality.

In other words, the driving force of evolution is the selective pressure acting on modified replicas of living organisms. The greater or lesser selective pressure to which organisms are subjected determines the rate of evolution, i.e., the speed at which new efficient designs for entropy dissipation are generated.

Information theory [24]

Although there isn't much more we can say about the thermodynamics of the behavior of living beings that is truly relevant, we are left with the challenging task of expressing and applying Schrödinger's abstract theorem in a concrete and useful way. We will begin this development by introducing Shannon's Communication Theory (1949), which provides us with a very useful tool for applying Schrödinger's theorem to the study of human behavior.

---

[24] The set of texts in this unit is a compilation of texts written between 1998 and 2003. (Editor's Note)

Shannon's theory

Although Shannon's communication theory is a mathematical work, I have included its reference in this section due to its capacity to translate the abstract concepts and principles of thermodynamics into a more understandable form for the study of human behavior. However, we must clarify the following:

1. Communication theory is a mathematical theory, so it does not provide any knowledge about the world but only about certain properties that the world and its objects may or may not possess (Bunge, 1960; Winner 1986).

2. Communication theory provides us with a language that allows us to express some terms, principles, and theorems of thermodynamics and thus facilitate their application to the field of human behavior.

3. Some authors have argued that there is no empirical dependency or causality relationship between Boltzmann's entropy and Shannon's uncertainty, with only a mathematical identity relationship existing between them. However, it seems reasonable to me to think that such a relationship does exist, as defended by other authors (Szilárd, 1929; Brillouin, 1953; Jaynes, 1957; Landauer, 1961).[25]

Of course, the mathematical theory of communication cannot add anything to what has been previously explained since it is only a new language. What we are trying to do is translate the factual content

---

[25] Leo Szilard proposed a thought experiment to demonstrate that the possession of a single Shannon bit of information indeed corresponds to a reduction in the entropy of the physical system. This idea has been recently demonstrated (Toyabe, et al., 2010), as well as Landauer's principle, which established that any information with a physical representation must somehow be embedded in the statistical-mechanical degrees of freedom of a physical system (Bérut, et al., 2012). Furthermore, Léon Brillouin derived a general equation stating that a change in the value of one bit of information requires at least $k \cdot T \cdot \ln 2$ energy. In his book, Brillouin further explored this issue and concluded that any change in the value of bits (measurement, yes/no decision, erasure, display, etc.) would require the same amount of energy. Consequently, according to Brillouin, acquiring information results in a reduction of local entropy and, at the same time, an increase in the thermodynamic entropy of the environment. (Editor's Note)

presented in the thermodynamic view into another language that is much more related to the processes we subjectively experience.

The theory of communication is based on and revolves around the concept of "uncertainty," defined as the amount of ignorance one has about "something." The mathematical function, H, that Shannon chose to define the concept of uncertainty was exactly the same one that Boltzmann chose to define the concept of entropy almost eighty years earlier.

"In 1961, one of the authors (Tribus) asked Shannon what he had thought once his famous measure was confirmed. Shannon replied: 'My biggest concern was what to call it. I thought about calling it 'information,' but the word was too common, so I decided to call it 'uncertainty.' When discussing it with John Von Neumann, he gave me a better idea. Von Neumann said: 'You should call it entropy, for two reasons: first, because your uncertainty function has already been used in statistical mechanics with that name, so it already has a name; second, and more importantly, nobody truly knows what entropy is, so in a debate, you will always have an advantage.'" (p. 253, in Tribus, M., McIrvine, E.C., Energy and Information. In Scientific American. "Energy". Madrid: Alianza Editorial. 1982)

In Shannon's communication theory, the objects of our knowledge/ignorance are mathematically represented by random variables (discrete or continuous), that is, by sets of events for which we can define a probability distribution.

Given a discrete random variable X defined by,

$$X = \left\{x_1, x_2, x_3, \dots x_n\right\} \text{ in which } \sum_{i=1}^{n} P(x_i) = 1,$$

the function H is defined as:

$$H(X) \equiv - \sum_{i=1}^{n} P(x_i) log P(x_i) \qquad (28)$$

Boltzmann discovered that entropy could be expressed using this function,

$$S = kH \qquad\qquad (29)$$

Where k is the Boltzmann constant.

On the other hand, Shannon saw in this function the ideal way to express his concept of uncertainty, that is, the measure of ignorance about X,

$$\text{Uncertainty about } X \equiv H(X)$$

and through it, define his concept of information as "reduction of uncertainty,"

$$I =- \Delta H \qquad\qquad (30)$$

It is also appropriate to introduce the concept of disinformation as negative information, that is,

$$Disinformation = - I = \Delta H. \qquad (31)$$

Furthermore, the information provided by a specific message (event, stimulus, etc.) to a particular receiver depends on both the message itself and the receiver. Thus, the same message can provide information to receiver A and disinformation to another receiver B, depending on their respective uncertainties before receiving the message.

These are the two fundamental concepts of information theory. However, since Shannon used the function that Boltzmann discovered as the statistical equivalent measure of entropy, that is,

$$S = kH \qquad\qquad (29)$$

where H is (28), there is controversy over whether entropy and uncertainty can be identified in practice. This depends on whether Thermodynamics can be extended to the realm of information. In other words, whether we can use the laws of Thermodynamics to understand phenomena in human communication.

Uncertainty and entropy

"Quantities of the form $H = -\sum_i p_i \log p_i$ (where K only represents the factor corresponding to the chosen unit of measurement) play a fundamental role in this theory as measures of information, choice, and uncertainty. The form of H is nothing other than that of entropy defined in certain formulations of statistical mechanics, where p_i is the probability of the system being in a particular state within its phase space. This H is then the H of Boltzmann's famous theorem. We will refer to $H = -\sum_i p_i \log p_i$ as the entropy of the probability set. If x is a random variable, we will write H(x) for its entropy, to distinguish it, for example, from H(y), the entropy of random variable y." (Shannon, 1949)

In statistical mechanics, entropy designates the degree of molecular disorder in a system, whereas in information theory, uncertainty designates the degree of disorder in the source of information. Are these two properties the same thing? This is the fundamental question for applying thermodynamic principles to the explanation of human nature.

What is uncertainty? I believe we would all agree that it is a subjective experience we all go through in every moment. Of course, it is a variable experience; sometimes we experience a lot of uncertainty, and other times we do not. Most of the time, we are not even aware of the uncertainty we experience. In essence, we can say it's an experiential fact.

When we colloquially say that we experience uncertainty, are we ultimately referring to an increase in molecular disorder somewhere in our body? Is the uncertainty we experience of a physical nature, or is it imaginary and immaterial? For those who believe that the experienced uncertainty has nothing to do with the physical processes occurring in their bodies, then there is no connection between uncertainty and entropy. However, if we admit that our subjective experience arises exclusively as a consequence of certain physical processes in our bodies, as has been extensively demonstrated, then we must conclude that the experienced uncertainty is the result of an increase in the entropy of specific physical processes in our bodies.

Let's take it step by step:

1. From a physical standpoint, our body has a certain level of entropy (molecular disorder), although it may not be technically measurable.
2. Our subjective experience depends solely (it is the result) on the physical processes in our body.
3. We experience something we call 'uncertainty,' meaning that uncertainty is an aspect of our experience, much like we experience cold or heat, joy or sorrow, etc.
4. The uncertainty defined in information theory (which accurately describes our experience of uncertainty) is mathematically identical to the physical entropy that accurately describes the molecular disorder of physical systems.

Therefore, it is reasonable to deduce from the above that the experienced uncertainty depends solely (it is the result) on the physical processes occurring in our body, which either increase or decrease the entropy of the organism.

When we refer to the experienced uncertainty, we are simply and unequivocally referring to the physical (molecular) entropy of our body, and consequently, the experienced uncertainty must obey the laws of thermodynamics, as it is nothing more than the correlation of the entropy of our body. Similarly, since information is the reduction of uncertainty, when we experience an increase in our information, when we acquire information, it means that there is a decrease in the entropy of our body's systems. Our body has molecularly organized itself.

Principle of psychophysical isomorphism

The scientific treatment of subjective experience as a real phenomenon can only be approached from a naturalistic perspective. Subjective experience, emotions, feelings, thoughts, and so on, are events produced by the chemical-electrical activity of the brain, which we can observe from the outside, using instruments, or from the inside, through our own experience. Our experience is the faithful reflection of the chemical-electrical activity of our brain.

In the 1930s, a small group of German naturalistic psychologists Wertheimer (1923), Kohler (1947), and Koffka (1935) proposed an idea that we can only fully understand today. They suggested that, in the absence of other knowledge, the relationship between neuronal activity and the resulting experience should be isomorphic. More specifically, the order of neuronal processes should be the same as the order of subjective experience. At that time, the mathematical theory of information had not yet been formulated, and this idea was not adequately understood.

Although Shannon was American and unaware of this idea, the formulation of information theory actually rests on this principle by identifying uncertainty (experienced disorder) with entropy (objective disorder).

Simplifying the chain of processes, we can say that objective disorder produces neuronal disorder that is experienced as uncertainty. Therefore, the order (disorder) of neuronal processes is equal to the order (disorder) of the corresponding subjective experience, that is, to the certainty (uncertainty) experienced.

What this principle asserts is that the experience of uncertainty corresponds to disordered chemical-electrical neuronal processes. In short and clearly, the entropy of neuronal processes is equal to the experienced uncertainty.

Therefore, an information process, that is, a reduction of uncertainty, consists of a physical process of reducing neuronal entropy, and vice versa.

I believe that this identity corresponds, in some way, to the facts we know, and therefore, I accept it as true even though I cannot provide laboratory evidence.[26] Consequently, I not only see no linguistic difficulty in identifying entropy and uncertainty but also see no difficulty in identifying them as equal facts or two sides of the same fact (objective and subjective).

Uncertainty is the subjective facet of our experience, and entropy is the objective facet of it, i.e., of the physical processes within our body (primarily our nervous system).

The mathematical theory of information and the principle of psychophysical isomorphism allow us to interchange our external and internal viewpoints of the same physical process. We can describe processes from the outside as entropy or disorder dissipation processes, or from the inside (subjective experience) as uncertainty dissipation or information processes.

Indeed, now we can reframe our understanding of dissipative systems from the perspective of subjective experience. We have mentioned that all living beings are characterized by their capacity for self-organization, that is, their ability to acquire structure and order. Therefore, we can also say that all living beings are characterized by their capacity for self-information, meaning their ability to reduce their uncertainty and insecurity.

A living being is a highly organized system, in other words, highly informed and secure. An increase in disorganization or entropy poses a threat to existence, meaning that an increase in uncertainty signifies danger.

---

[26] As we mentioned earlier, this relationship between uncertainty and entropy has been studied (Szilárd, 1929; Brillouin, 1953; Landauer, 1961) and demonstrated (Bérut et al., 2012; Toyabe et al., 2010) in recent years. (Editor's Note)

Objective experience	Subjective experience
Entropy, disorder	Uncertainty, insecurity
Self-organization	Self-information

Experienced uncertainty, that is, the entropy or disorder of neuronal processes, manifests itself in various forms such as anxiety, fear, oppression, insecurity, restlessness, or concern, in other words, as negative experiences. On the contrary, the experience of certainty manifests as satisfaction, pleasure, happiness, etc.

This equivalence between entropy and experienced uncertainty allows us to simultaneously analyze objective physical behavior and its respective subjective experiences. So now, when we seek to understand human behavior, we can equally understand their subjective experiences.

In many instances, we cannot directly observe human behavior, and instead, we can learn about subjective experiences through interviews. Since our goal is to understand human behavior, this expression, through information theory, of the functioning of living beings seems very relevant and constructive.

From experience, we know that we are interested in information and that we disdain disinformation and uncertainty. We also know, through experience, that the information we possess is not easily preserved, and we must work to keep it that way. We know that we can acquire information as long as we put in a lot of effort. We also know that acquiring disinformation, that is, experiencing an increase in our uncertainty about ourselves and the world around us, is very easy.

Of course, there is a great conceptual confusion in identifying 'message' with 'information.' Most people think that everything is information, that any news brings information to them. But this is not the case at all. A message only brings information when it is capable of reducing our uncertainty, and this does not happen very often. Most of the messages we receive bring us disinformation because they increase our uncertainty.

Apply information theory

Substituting the term 'entropy' with 'uncertainty' and replacing 'reduction of entropy' with 'information' in the propositions of thermodynamics provides us with a new perspective on the behavior of living beings. In particular, it allows us to understand the subjective experience associated with objective physical behavior.

- If in the thermodynamic formulation, we stated that living systems are capable of staying far from thermodynamic equilibrium, using the language of information theory, we can say that they are capable of staying far from uncertainty and, therefore, they remain extremely informed.
- Therefore, life depends on the ability to stay informed. To achieve this, constant interaction with the environment is necessary to eliminate the uncertainty that inevitably arises in living organisms.
- In this sense, uncertainty is something negative and undesirable, while moving away from or reducing uncertainty is something positive and desired by living beings.
- Schrödinger's theorem states that living beings obtain information from their environment in an amount less than the disinformation they produce in it. In other words, to reduce our uncertainty, we must increase it to a greater extent in our environment, or in other words, information is never free, as is commonly and erroneously thought today.
- The main sources of information for living beings are other living beings in their environment, which can lead to a competition for information among them.

These formulations of certain thermodynamic propositions in terms of information theory are only really useful in the case of human behavior, as the development of cultural life cannot be addressed, at least for now, using classical thermodynamic terms.

Likewise, the concept of entropy cannot be applied to our subjective experience, whereas the opposite is true for the concepts of uncertainty and information. For example, if someone loses their job, we can understand that their uncertainty has increased, while it is very difficult to understand that their entropy has increased, even though in reality, it's the same thing; language has these peculiarities.

The importance of this perspective is that it allows us to understand that real information, not the mathematical concept of information but what exists in actual facts, is not an immaterial fact but quite the opposite. Obtaining real information means producing entropy in another system, as obtaining real information is nothing more than reducing one's own entropy and requires expending energy. For example, a biologist needs to dissect (disinform) the corpse of an animal to learn about what it recently ate. Thus, we arrive at the paradoxical result that it is not possible to acquire knowledge without producing entropy in other systems, meaning that all knowledge, all learning, etc., comes at a high price, and someone has to pay it. These are not immaterial processes, and therefore, they are subject to the strict economy of nature. Educators must begin to understand that the acquisition of any form of knowledge can only occur in exchange for the expenditure of energy or other types of resources (emotional, financial, or health-related) from other systems (teachers, parents, non-adaptive children, etc.).

The behavior of living beings [27]

Thermodynamically, we have seen that life depends on living beings staying far from their thermodynamic equilibrium. To achieve this, they must be capable of engaging in interaction processes with their environment in which they increase the entropy of the environment to a greater extent than they reduce their own entropy (17).

As we have observed, this limitation results in a competition for entropy among living beings since entropy can only be dissipated in highly ordered systems, namely, the other living beings in the environment. Thus, we can state the first and most important principle of the behavior of living beings:

No matter how complex and varied the behavior of living beings may be, it always consists of success or failure in the transfer of their entropy to their environment.

In other words, when faced with the question:
"What is that living being doing now?"
The answer should always be something that means:
It is either succeeding or failing in transferring its entropy to the environment.

The success or failure of each living being in performing this function will determine its survival or its illness and death, within the endless cycle of life.

---

[27]The texts in this unit are a compilation of various texts, but they are primarily taken from the document "Scientific Foundations of Human Behavior" from the year 2007 and from the letters exchanged between Dr. Esteve Barrull and Professor K.H. Norwich between 1998 and 2004. (Editor's Note)

The perception-action cycle in animals

The nervous system is an organ unique to the animal kingdom, and its primary function is to direct the motor capabilities of animals to keep them adapted to specific circumstances, allowing them to survive and reproduce. How does the nervous system perform this function? How does it successfully guide the behavior of animals, avoiding dangers, finding food, or caring for offspring? Succinctly, it does so by acquiring and storing vital information for the organism.

In this section, I will explain the theory of the perception-action cycle that I developed between 1993 and 1998, and had the opportunity to refine through the work I conducted with Professor Kenneth Norwich from 1998 to 2004.[28]

---

[28] The study of the nervous system within the framework of thermodynamics linked to information theory is a novel perspective, but some authors have already begun to work in this view. Some of them have precisely modeled brain activity in thermodynamic terms (La Cerra, 2003; Varpula et al., 2013), and cognitive processes have been modeled in terms of information (Anderson, 1996; Friston, 2010), as well as in both perspectives simultaneously (Collell & Fauquet, 2015). Furthermore, this perspective of information entropy has been used as a framework for studying psychological phenomena in various research studies (Barton, 1994; Carver & Scheier, 2002; Hollis, Kloos, & Van Orden, 2009; Vallacher, Read, & Nowak, 2002; Hirish, Mar & Peterson 2012). For example, researchers have observed self-organization dynamics during the problem-solving process (Stephen, Boncoddo, Magnuson & Dixon, 2009; Stephen, Dixon & Isenhower, 2009), suggesting that cognitive-behavioral systems follow the same basic principles as other dissipative systems. On the other hand, similar interpretative frameworks have been applied to understand the neural substrates of cognitive operations. In particular, several techniques have been developed to quantify entropy levels within neural systems (Borst y Theunissen, 1999; Nemenman, Bialek & Ruyter van Steveninck, 2004; Paninski, 2003; Pereda, Quiroga y Bhattacharya, 2005; Strong, Koberle, de Ruyter van Steveninck & Bialek, 1998; Tononi, Sporns & Edelman, 1994). Finally, Karl Friston seems to have arrived at a similar conclusion to Esteve's and has explicitly emphasized the importance of entropy minimization as an organizing principle of neuronal function (Friston, 2009, 2010; Friston, Kilner & Harrison, 2006). (Editor's Note)

Perception as a source of uncertainty

After realizing that there was a way to build an understanding of the objective and subjective experiences of human behavior through thermodynamics and information theory, based on Schrödinger's principle, I searched for articles that could relate to or contribute to this viewpoint. I found very few articles interested in thermodynamics and the behavior of living beings, and all of them approached this topic in a very general and diffuse manner. However, I found a small treasure, a very rare find in today's times. It was the work of Kenneth Norwich (1993), who had discovered the general function of the response of sensory neurons, which I called the "Norwich Law." The importance of this function was not only to consider all the known empirical functions of the sensory neuron response but also to take into account the entire response of sensory neurons over time, i.e., the phenomena called "adaptation." Furthermore, my surprise was to find that the Norwich Law confirmed my viewpoint on thermodynamics as a way of understanding the behavior of living beings.

Neurologists had accumulated a set of different response functions of sensory neurons (under constant stimulus), and each of them worked in different situations (Fechner, 1850; Stevens, 1957; Weber, 1846). The response of the sensory neuron is well-known and characterized by a negative exponential function (Laughlin, 1989). When a constant stimulus hits a sensory neuron, it starts with a high peak of electrical wave trains that exponentially decrease until reaching a null response after some time.

Neurologists had theorized that the intensity of the initial response was proportional to the intensity of the stimulus, which was confirmed by empirical data (Stevens, 1957). However, they could not find a neurological function for the subsequent decrease in neuronal response even when the stimulus intensity remained constant, which is called the "adaptation period." The general explanation for this phenomenon was that the sensory neuron exhausts its capacity to transmit electrical waves (Desimone & Duncan, 1995; Kandel, Schwartz & Jessell,

2000). In other words, to the question, "Why does a sensory neuron, in response to a constant stimulus, progressively stop responding?" Neurologists could only answer, "Because it's tired."

Norwich unified the set of known empirical functions of the sensory neuron with the following function:

$$F(t) = k \cdot H_{Shannon}(stimulus) \qquad (32)$$

where F is the intensity of the sensory neuron response (or subjective intensity of the stimulus), k is a constant that accounts for units, and H is the Shannon uncertainty function of the constant stimulus. In this way, he discovered that instead of responding to the intensity of the stimulus, which remains constant over time, the sensory neuron responds to its uncertainty about the stimulus, which decreases over time.

Norwich hypothesized that when a stimulus hits a sensory cell, it begins to sample the intensity of that stimulus at a constant frequency. The sensory cell does not notify the sensory neuron of the value of the stimulus intensity, but rather the uncertainty about it. In other words, the sensory cell tells the sensory neuron how much uncertainty there is about the stimulus intensity, not the stimulus intensity itself.

If the stimulus intensity is constant, initially, the uncertainty is maximum because there is only one sample. But the uncertainty decreases exponentially as the sensory cell takes more samples.

Norwich modeled $F(t) = k \cdot H_{Shannon}(stimulus)$ (32) for a constant stimulus to find that

$$H_{Shannon}(stimulus) = \frac{1}{2}\ln(1 + \beta I^n/t) \qquad (33)$$

that is to say,

$$F(t) = \frac{1}{2}k \cdot \ln(1 + \beta I^n/t) \qquad (34)$$

121

For this case, Norwich's law (32) became a negative exponential function that encompasses the entire sensory neuron response process, including the adaptation process.

With his law, he made the most significant theoretical improvement in neurology since the time of Weber (1846) and Fechner (1850) by linking the behavior of sensory neurons to information theory. If neural systems were understood as mechanisms for processing information, this theoretical connection would become fundamental.

Thanks to his work, we know that animals only perceive uncertainty. Whenever a stimulus is completely known, it ceases to be perceived. The intensity of stimulus perception is proportional to its uncertainty, which, in the initial moment, is proportional to the stimulus intensity[29], as established by the early neurologists. However, if the stimulus remains constant, the uncertainty about its intensity is reduced through continuous sampling by the sensory cell. When uncertainty is completely reduced, perception ceases.[30] The job of sensory cells is to gather information about the stimulus intensity through continuous sampling of the stimulus intensity, but this information is not transmitted to the sensory neuron. Instead, the sensory cell transmits the uncertainty of its measurement.[31]

Con todo ello podemos enunciar el primer axioma del principio de percepción-acción que gobierna el comportamiento de los animales:

---

[29] It is well-known in physics that the variance of a measurement is proportional to that measurement. Measuring the distance between two cities in kilometers will have a variance measured in meters, while measuring the thickness of window glass in millimeters will have a variance measured in at least 1/10 millimeters.

[30] It doesn't mean that uncertainty has become zero but rather it is equal to a characteristic reference uncertainty for each type of sensory receptor. In other words, a sensory receptor can be understood as a measuring device that has its own variance limits. For example, with a common tape measure, it's not possible to measure 1/100 millimeters.

[31] Allow me to put it in human terms: the sensory cell doesn't tell the sensory neuron, "my current estimate of the stimulus intensity is 14 units of intensity," but rather "my current uncertainty about the stimulus intensity is 14 bits."

Axiom 1: *The intensity of sensation or perception is measured by the amount of uncertainty associated with the stimulus.*

The fact that sensory neurons respond to the entropy (uncertainty) of stimuli allows us to understand the neuronal function within the framework of Schrödinger's theorem for living beings. The perceptual system is a channel through which uncertainty (entropy) is transmitted from the environment to the central nervous system. In other words, we only perceive uncertainty. In a constant stimulus field with no uncertainty, there would be no perception at all. Therefore, every perception entails an increase in uncertainty or entropy in the central nervous system, in other words, an increase in the circulating electrical energy.

Our observation of perceptual behavior completely confirms this explanation. Every organism with a neuronal perceptual system directs its attention to unknown, uncertain, or more disordered stimuli. When an accident happens on the highway, vehicles traveling in the opposite direction also stop, causing another traffic jam, because the drivers direct their attention to the most disordered part of their visual field (the accident). This results in a flow of disorder (uncertainty, misinformation) through the drivers' perceptual system, from the accident to the vehicles traveling in the opposite direction.[32]

Marketing and propaganda strategies exploit this situation by designing very chaotic advertisements to capture the attention of potential viewers. In reality, 'good' propaganda is the one that generates the maximum uncertainty in viewers: "you will be the best," "you will succeed," "the bread that doesn't make you gain weight," etc., in other words, messages that disrupt the status quo.

Valued artistic and scientific production is the one that disrupts the intellectual field of viewers, the one that irritates and even produces

---

[32] This law of perception developed by Norwich (1993) is consistent with the free energy theory developed by Karl Friston and his colleagues (2009, 2010), who have proposed the idea that attention is directed toward stimuli that are uncertain or unknown, in other words, those that increase the entropy of the nervous system. (Editor's Note)

aversion. Beethoven, Picasso, Einstein, or Kandinsky created 'stimuli' that irritated people of their generation, meaning that they drew their attention to their work. In conclusion, a stimulus that does not generate uncertainty is not perceived.

What we experience is not what we perceive

One of the difficulties in accepting Norwich's law is that you might think your experience is always complete, not just limited to what is uncertain. In other words, you see all the things in your visual field, regardless of whether they are known or unknown.

Norwich's law states that we only perceive stimuli that are uncertain to us, and the intensity of this perception is proportional to the uncertainty of their measurement. This means that we do not perceive all stimuli that are certain (have no uncertainty), which seems to contradict our experience.

This phenomenon can be explained by distinguishing between perception and experience. That is, considering perception as all the messages coming from our sensors and experience as all the messages that are currently active in our neural system, which includes all the messages from our perception and our memory. From this perspective, our experience would be a blend of perception and memory. From perception come all the stimuli that are uncertain, and from memory come all the stimuli that are certain. The complete image of what we see is an active composition of what we perceive with our eyes and what we store in our current visual memory. All familiar thoughts come from our memory, and only unknown or new events come from our perception.

Even though all the things in our home are known to us, some of them are in different places over time. Our perception accounts for what is uncertain to us, i.e., the location of those objects, while memory accounts for what they are. I know what my TV remote looks like because I've had it for many years and have used it thousands of times. But generally, I don't know where it is, either because I have trouble remembering where I left it or because other people have moved it. So when I see it, even though I see the complete remote, only its position comes from my perception (from my eyes), while the shape I see comes from my memory.

Perception economy

However, I can look at my TV remote and perceive new things about it that are uncertain to me. These new things might be small details that I didn't notice before. In other words, even for the most familiar things, there are always small amounts of uncertainty left to perceive. This is due to our way of perceiving and storing information in our memory.

One way to justify why we only perceive uncertainty is because it's the most economical way to perceive and store information. I realized this after studying the mathematical methods of factor analysis used in statistics and quantum mechanics.

I assume you are familiar with the basic concepts of factor analysis, as they are too extensive to explain here. If you are not familiar with them, it may be difficult to understand this topic.

Let me explain this problem in a simplified way. Suppose you can only see in grayscale. Then, what you see at each moment can be represented by an n x n dimensional matrix of "pixels," each of which can take values from 0 to 255, representing different shades of gray. This matrix can be considered the raw material of visual stimuli at any moment. Since we have good visual sensitivity, each side of that matrix could be around 10 thousand pixels. Our visual matrix could have a size of around 100 GB. If we were to save the entire matrix, we would need to store 100 GB every moment, which seems very disproportionate or uneconomical.

Instead of storing the entire matrix, it seems reasonable to store only the most important features that appear in that visual field. For example, if a venomous snake appears in front of us, it doesn't matter if it has some specific detail.

So, what are the most important objects in each visual matrix? The ones that are more uncertain will be the most important because, as living beings, we stay alive by avoiding entropy. In other words, we need to find (perceive) the most uncertain elements in that visual matrix.

To do this, our brain performs factor analysis, looking for dimensions with the most uncertainty (variance in statistical terms). Only these dimensions are stored in our brain, discarding the rest of the less uncertain dimensions. By working in this way, we can perceive and store 80% of the uncertainty with just a 5 x n matrix, for example.

The perception-action cycle as a dissipative process

The Norwich law suggested to me a clear understanding of the behavior of neuronal systems and their contribution to the general law of behavior, which is to keep living beings away from their thermodynamic equilibrium.

We know that living beings need to continuously transfer their entropy to the environment, or in other words, extract order from their surroundings. We have discussed earlier that the best systems for dissipating entropy are those with high relative order, meaning those that are far from their thermodynamic equilibrium. So it is clear that the neuronal systems of animals must be designed to help them find these places. Therefore, a living system must be capable of moving away from sources of entropy in its environment and moving closer to sources of order.

Theoretically, orienting the behavior of animals towards survival could be achieved in two different ways:

1) through a mechanism that is sensitive to anything that represents a resource for survival (food, territory, mate, water, safety, etc.) and directs the organism towards these goals.

2) through a mechanism that is sensitive to any danger to survival (predators, hunger, thirst, loneliness, etc.) and causes the organism to flee from these situations.

Thanks to Norwich's discovery, we know that the nervous system functions as a mechanism of the second type, that is, as a mechanism that detects dangers to survival and directs the movement of animals to escape from them.

The perceptual system acts as an alarm system. We have seen that perception is proportional to the uncertainty of the stimulus and that the afferent nervous system transduces the stimulus uncertainty into electrical charges that flow towards the nervous system. In this sense, we see that the sensory cell is an 'entropy transducer' that converts the entropy of the stimulus into electrical signals that are carried by the sensory neuron to the brain. The longer the sensory cell takes to reduce its uncertainty, the more entropy or uncertainty will reach the brain.

Therefore, the total electric charge flowing in the nervous system is a measure of the uncertainty experienced by the subject. According to Schrödinger's theorem, living beings need to keep their relative entropy extremely low to stay alive, and that is precisely the function of the efferent part of the nervous system. When the electric charges reach the efferent part, they end up being transmitted to the glands and muscles, which produces the organism's action. The organism, by acting, changes the stimulus situation, either by increasing or decreasing the perceived uncertainty.

Indeed, since perception only results in an increase in neuronal entropy and experienced uncertainty, there must be mechanisms to reduce it, either towards other subsystems of the organism or towards the external environment. This means that the dissipation functions need to be activated to reduce the entropy acquired in perception.

In conditions of safety (order), living beings have no need to act. Only when there is an increase in their entropy do they need to act to reduce it. Therefore, the production of the electrical energy necessary to activate behavior can only be generated in response to the entropy received by the organism. The greater the entropy it perceives, the greater the generation of electrical impulses in the perceptual system, and hence the greater the behavior generated, which will attempt to eliminate the acquired entropy. We can see how the nervous system has evolved to more efficiently dissipate entropy.

If it were the opposite, that the perceptual system generated electrical energy proportionally to the order received by the organism, when it received disorder, it would not produce electrical energy and therefore would not be able to act to reduce it, which would quickly lead to the death of the organism.

In summary, neuronal systems can be understood as simple entropy dissipation systems in the following way:

1) Through perceptual devices (sensory cells and sensory neurons), the brain obtains perceived uncertainty as electrical charges (via the afferent system).

2) Through the efferent system, the brain reduces its uncertainty by sending it to action devices (muscles and glands) as electrical charges.

To the 'law of Norwich,' which states that the intensity of perception is measured by the amount of uncertainty associated with the stimulus, we can add a second axiom that states:

Axiom 2: *The perceiver always acts in such a way as to reduce uncertainty and, consequently, reduce the intensity of sensation or perception.*

This axiom completes the perception-action cycle in the process of acquiring information. The perceiver takes an active role in the process of perception to reduce uncertainty; they cannot be a passive receiver of information. Action is required, energy must be expended, and effort must be made to obtain information.

The simplest example of this phenomenon can be found in sensory cells. As discovered by Norwich, sensory cells sample the stimulus. If the stimulus is constant, each sample reduces the uncertainty of the stimulus's intensity. Thus, within the sensory cell, both perception and action occur simultaneously, as it evaluates the stimulus intensity and takes new samples while reducing uncertainty. In other words, sampling (action) is the way the sensory cell reduces perceived uncertainty. If the sensory cell couldn't take samples, it also couldn't reduce perceived uncertainty. Only with a very long sample can uncertainty not be reduced. Consequently, without some form of action, we cannot reduce perceived uncertainty, i.e., we cannot obtain information.

The principle of perception-action, as stated here, is one of many similar laws in nature where the initiation of a process leads to a compensatory process that opposes the first. Lenz's Law (1834) in electromagnetic theory states that if an external magnetic field is applied to a conductor to change the number of magnetic flux lines passing through it, an electric current is induced in the conductor whose associated magnetic field opposes the change. In simpler terms, an applied magnetic field induces a process that tends to reduce the magnetic field. Lenz's Law is a specific case of Le Chatelier's Principle (1888), which states that whenever an external disturbance is applied to a system in equilibrium, the equilibrium shifts in a way that will reduce the effect of the disturbance. In the case at hand, whenever a sensory stimulus is applied to a perceptual system, the system responds in such a way as to reduce the effect of the sensory stimulus. In other words, sensory adaptation, in its broadest sense, embodies Le Chatelier's Principle (Norwich, 2010).

One of the most important issues in analyzing this second axiom is the generalization of the term "adaptation." Classically, adaptation refers to central or sensory physiological processes that tend to diminish the effect of the stimulus and can be represented at two levels: psychophysical and neurophysiological. Psychophysically, adaptation is associated with a reduction in the subjective intensity of a stimulus: for example, the intensity of an odor tends to decrease over time, sometimes becoming imperceptible. Neurophysiologically, the rate of impulses in an afferent neuron decreases over time after the application of a stimulus. Furthermore, thresholds tend to increase as a consequence of adaptation. In the context of the theory I am presenting to you, the concept of adaptation is broadened to include any activity, at any level of complexity, that tends to reduce the biological effect of an applied stimulus. Therefore, moving away from a stimulus by any means (avoiding the eyes from a bright light, removing your hand from a flame, etc.) should be considered a process of adaptation.

It may seem contradictory that the reduction of the sensory effect produced by the action of voluntary muscles should be included in the expanded definition of adaptation. However, I believe that the result of doing so is to admit a generalized Le Chatelier principle in the field of perception. One consequence of this second axiom is that a sensory system should be considered together with its active or motor component. Perception is not a passive process. The eye is more than a sophisticated camera that absorbs and analyzes light signals. Rather, the eye acts with biochemical adaptation mechanisms, along with extraocular muscles, and in principle, other muscles, both under reflex and voluntary control, to facilitate an overall process of adaptation.

In its entirety, we thus see a more complete picture of the neuronal process, which I have called the principle of perception-action, and which establishes that perception electrically charges the brain (increasing its uncertainty) while action discharges it (decreasing its uncertainty). This cyclic process of adapting to perceived uncertainty through action is the elementary mechanism of animal behavior. As you can imagine, this process shows us that the nervous system is a dissipative system that aims to maintain its state of minimum entropy.[33]

---

[33] Recently, some authors like Karl Friston have reached the same conclusion that Esteve and Norwich arrived at in the late 1990s. According to them, the primary goal of any nervous system is to minimize the experience of entropy by continuously modifying neural structures in response to environmental information (Friston, 2009, 2010). (Editor's note)

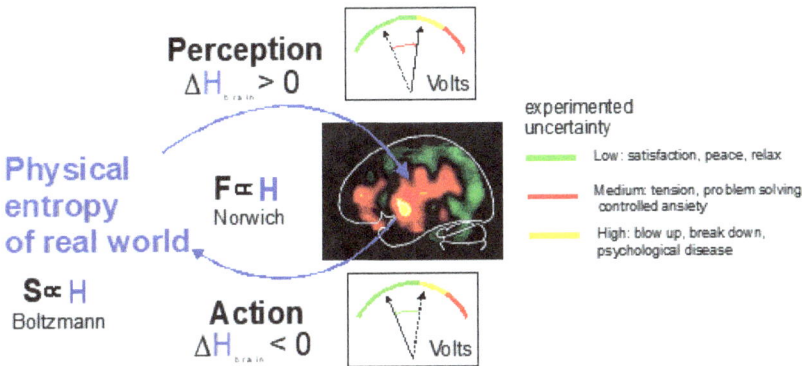

Figure 5. Original drawing created by Esteve Barrull in 1998 in the first letter sent to Professor Norwich to explain the perception-action cycle. In the image, we can see that the physical entropy of the world is transferred to the brain through the perception of uncertainty in stimuli, resulting in an increase in neuronal entropy and electromagnetic activity of the nervous system. Through action, generated by the electromagnetic activity of the nervous system, the perceiver attempts to reduce the perceived uncertainty in what we can call a process of dissipating the entropy acquired through perception.

Allow me to describe a very simple case. Let's assume a completely discharged neuronal system, with no activity - no perception and no action. At some point, a constant stimulus hits a sensory cell. Right at that moment, a stream of electric charges flows towards the brain. However, as the brain receives electric charges, they are redirected to the muscles and glands, producing some kind of response (movement). As a result of the action, the perception of the stimulus changes.

The change in perception caused by the action of the neuronal system can be an increase or decrease in the uncertainty of the stimulus. If the uncertainty of the stimulus increases, more electric charges will flow to the brain and more charges to the muscles and glands, so the action will also increase. Conversely, if the uncertainty of the stimulus decreases, fewer electric charges will flow to the brain and fewer charges to the muscles and glands, so the action will also decrease.

In other words, when an increase in perceived uncertainty occurs, there is also an increase in action, which increases the probability that the organism will perform the required action to reduce uncertainty. On the contrary, if action succeeds in reducing the perceived uncertainty, there will also be a decrease in the subsequent action, and the organism will tend to stay near this state. In conclusion, the action of neuronal systems is proportional to their perception. This means that perceiving more uncertainty results in a higher intensity of response, and vice versa. The outcome of this system is that neuronal systems always lead living beings to remain in places where uncertainty is minimal.[34]

In summary, I have attempted to analyze the general process of behavior in two phases, each defined by a single axiom. The first phase quantifies perception by estimating its uncertainty value. In simple cases, the intensity of perception can be calculated using a standard weighted sum of probabilities to provide a Shannon entropy. In more complex cases, we must be content with a more subjective assessment of intensity. The second phase is characterized by motor activity, designed either voluntarily or reflexively, to reduce the uncertainty or entropy from the first phase. In simple cases, the reduced uncertainty can be quantified as the information from perception. Thus, we have seen that this cycle of adaptation to uncertainty functions as a dissipative system that guides the organism to minimize its uncertainty.

With this, we have expanded the traditional meaning of the concept of "adaptation" to encompass any motor activity, whether voluntary or reflexive, that results in uncertainty reduction. Thus, a spinal reflex would, according to this definition, be considered an adaptive process. Therefore, the spinal reflex that governs the

---

[34] You may have noticed that throughout this process, I haven't talked about the direction of movement or the goal of the organism's action. That's because if we understand that any action is a response to a previous perception, even if the action is entirely random, the organism will continue to move until it achieves a reduction in perceived uncertainty. Therefore, no pre-set goal is needed to guide the mechanisms of action.

withdrawal of a hand from a hot object would be classified as an adaptive process. Similarly, a tendon stretch reflex can be viewed as a motor process that promotes adaptation by reducing the stretching in a tendon and muscle. Classifying a stretch reflex as an adaptive reflex is perhaps useful, but above all, it helps us see a stretch reflex as an example of the Le Chatelier principle: a stretched tendon shifts the equilibrium in a muscle, which then contracts to establish a new equilibrium, even if only briefly.[35]

It would be presumptuous to assume that all cases of perception could be placed under the umbrella of these two simple axioms. However, I believe they lead us in the right direction.

---

[35]A thermostat is a synthetic device that emulates the effect of the Le Chatelier principle, which seeks to establish the new equilibrium as close as possible to the displaced equilibrium. In fact, I realize that perception/action devices are systems for controlling the environment of living beings and not for controlling the living beings themselves. Any living being depends directly on its environment. So, by controlling their environment, they can obtain everything they need to survive. Living beings are the result of the work done by their environment on them. In the end, we are the result of the work of the Sun. In other words, it's not a behavior controller or a perception controller; it's an environment controller. I earnestly ask you to consider this idea. Think of it as adaptation failure also being a failure in environmental control, and vice versa; adaptation is succeeding in environmental control. This clearly connects us to Darwin's idea of adaptation to the environment.

Action as a source of information

While it is necessary to clearly separate both processes (perception and adaptation) to clarify their functions, they both operate simultaneously and are intrinsically linked, giving rise to a new phenomenon of their joint activity: the acquisition of information.

1) The asymmetric link

There is a fundamental link between both processes: perception processes provide the energy to activate adaptation processes. In other words, there is no action or adaptation without perception (axiom 2) and, ultimately, without uncertainty (axiom 1). The neural impulses that excite motor cells, such as muscle cells and gland cells, are triggered by the perception processes of sensory cells. As sensory cells generate more neural impulses, more actions or adaptations are carried out.

On the other hand, action processes may or may not succeed in reducing perceived uncertainty. If they succeed, adaptation takes place, and the perception of stimuli ceases. However, if they fail, adaptation is not achieved, and the perception of stimuli persists, leading to what we could call hypersensitivity (phobias, obsessions, etc.) and hyperactivity (stress, tics, aggression, etc.), which could be dangerous for the perceptive system. Success depends on the skills of the neural system and the characteristics of the situation.

2) Working together: acquiring information

As defined by Shannon, information is obtained when uncertainty is reduced. Therefore, the processes of perception and adaptation have evolved to allow neural systems to obtain information from their environment.

While perception always identifies where the source of maximum uncertainty is located (axiom 1), adaptation always works until perceived uncertainty is reduced (axiom 2). As a result, neural systems can acquire information.

Of course, each neural system is limited by its capabilities of perception and adaptation, meaning its capacity to deal with uncertainty. However, in relation to their capabilities, each neural system is designed to acquire information. Keep in mind that this theory does not assert that neural systems are designed to process information but to acquire information. The emphasis is on acquisition, not on information processing. Traditional theories claim that neural systems are designed to process information. In fact, if this were true, it would have no biological sense. What a living system really needs is to acquire information.[36]

However, a net gain of information is not guaranteed in all situations. Success in finding uncertainty and reducing it depends on the skills of the neural system and the characteristics of the situation. In many cases, the perception process cannot find the maximum source of uncertainty, or the adaptation process cannot reduce the uncertainty of the stimulus enough to stop generating more perceived uncertainty. Many circumstances can contribute to dysfunction in the neural system, but I won't address them now.

In conclusion, both processes, perception and adaptation, working together, serve a unique function: acquiring information from the neural system's environment.

---

[36] Acquiring information is another way of saying acquiring negentropy or reducing relative entropy, in other words, staying away from thermodynamic equilibrium. (Editor's note)

So, we can derive from Axioms 1 and 2 a first Corollary which states:

Corollary 1: *Neuronal systems are designed to acquire information from their environment by reducing perceived uncertainty through their motor or efferent activity.*

$$I_{acquired} = -\Delta H_{perceived} \tag{35}$$

The information acquired by a neuronal system is measured by the reduction of its perceived uncertainty. This result allows us to reformulate the second axiom in informational terms as follows: The perceptor always acts in such a way as to acquire information. Although this formulation cannot be understood without the appropriate formulation of axiom 2 in terms of the reduction of perceived uncertainty, it appears much more intuitive than its more technical formulation.

Emotions and the perception-action cycle

As we have seen, the nervous system plays a fundamental role in the process of dissipating the entropy of living beings:

1) Through perception, the brain acquires entropy:
   1.1) From the environment (external perception)
   1.2) From its own internal processes (internal perception)

2) Through action, the brain reduces its entropy:
   2.1) Transmitting it to other living beings in the environment
   2.2) Minimizing the perceived entropy

However, the nervous system also has emotions to guide animals toward survival. Even if you have a basic understanding of the sensory-neuro-motor scheme that describes the principle of perception-action, you will realize that it not only describes the logic of animal behavior but also its emotional states.

In the 1960s, Serbian biologist Vladimir Wukmir published what we could call the first comprehensible explanation of the nature of emotions. Wukmir (1960, 1967) established that emotions are the mechanism that living beings have to guide their behavior in the direction of survival and reproduction. For Wukmir, emotions are the perceptual systems that living beings possess to continuously assess the probability of survival presented to them in each stimulus situation. When the probabilities decrease, they experience negative emotions, and when they increase, positive ones. Emotion informs the organism about the favorability of each situation.

We would say, then, that emotion behaves as an intensive state variable (the total value is equal to the average of the parts). Each state of our organism corresponds to an emotion, which is more positive when it is a healthier, more life-oriented state (orexis) and more negative when our state approaches sickness and death (anorexis). Emotions are, therefore, the impulse for action (emovere) to find situations favorable to survival. Living beings try to move away from situations in which they experience negative emotions while trying to stay in those in which they experience positive emotions.

This naturalistic explanation aligns perfectly with the principle of perception-action that I have proposed.

Taking Wukmir's concept of emotion, which defines it as a biological mechanism that guides the behavior of living beings toward survival, the principle of perception-action provides a deeper explanation of the nature of emotions. Wukmir stated that emotions are manifestations of a measure of the survival adequacy of the perceived situation. If the situation is perceived as favorable for survival, the organism experiences a positive emotion, and if the situation is perceived as unfavorable, the organism experiences a negative emotion.[37]

If we equate "survival adequacy" with "reduced uncertainty," in other words, with "gained information," the principle of perception-action can explain emotional phenomena. Based on everything we have discussed about Schrödinger's Theorem and Norwich's Law, we know that life is directly related to the acquisition of information or the reduction of uncertainty, so this identification seems correct.

Consequently, based on the axioms of the principle of perception-action, we could establish that what emotions measure (the adequacy of survival in the perceived situation) indeed corresponds to the information obtained or the reduction of perceived uncertainty. If perceived uncertainty is reduced, we acquire information and

---

[37] Of course, this mechanism doesn't always work correctly and could fail in its assessments. In many cases, emotions may not be able to guide survival, as is the case with emotions in drug addicts. But we won't delve into the many circumstances that could interfere with the proper functioning of emotions for now.

experience positive emotions. If uncertainty is not reduced, we do not gain information and experience negative emotions.[38]

$$emotion \; = \; - \; \Delta H_{perceived} \; = \; I_{acquired} \qquad (36)$$

In this way, we can establish a second corollary of the principle of perception-action:

Corollary 2: *The reduction of uncertainty is manifested in high levels of neuronal organization through positive emotions such as the feeling of pleasure or satisfaction. Failure to reduce uncertainty is manifested by negative emotions such as the feeling of discomfort or pain.*

When a living being experiences an increase in its uncertainty (remember that uncertainty or entropy means death for life), the flow of electric charges in its nervous system increases, and consequently, the amount of action or behavior also increases, allowing the living being to move away from the situation. In this case, we can say that it has experienced negative emotions.

On the contrary, when it experiences a decrease in its uncertainty, the flow of electric charges in the nervous system decreases, and the amount of action or behavior also decreases, causing the living being to tend to stay in this situation. Then, we can say that it experiences positive emotions.

So, when we observe an animal running wildly or acting frantically, we can say that the sensory system of that animal is particularly excited, and therefore, it perceives a high level of

---

[38] I would like to clarify that any emotion is always a measure of the change in uncertainty and not the measure of uncertainty itself. This explains, for example, the relativity of emotion. People living in poor conditions may feel happy about things that make people living in good conditions feel bad. Even if uncertainty is high, any small reduction in uncertainty produces a positive emotion. Conversely, even if uncertainty is low, any small increase in uncertainty produces a negative emotion. This is why we can never be completely happy.

uncertainty in its sensory field (hence its high motor activity). In the first case, the animal experiences dissatisfaction, insecurity, restlessness, which means a lack of resources for survival and reproduction. In the second case, it's the opposite.

Then, we could reformulate Axiom 2 in emotional terms as follows: The perceiver acts, whenever possible, in a way that leads to experiencing positive emotions (which again seems much more intuitive than its more technical formulation).

Remember that, in accordance with Axiom 2, a generalized process of adaptation involving voluntary motor activity serves to reduce the uncertainty associated with the stimulus. We have supported the general principle that voluntary motor activity will always be aimed at promoting the perceiver's goals and, therefore, increasing the perceiver's pleasure or satisfaction. To put it more simply: if you do it on purpose, you do it because you like it. Therefore, we assert that the process of adaptation, particularly when carried out by voluntary means, will serve to enhance the well-being and, therefore, the "happiness" of the perceiver. Conversely, a lack of adaptation will result in a reduction in satisfaction or happiness and a sensation of pain. In other words, adaptation is associated with pleasure and a lack of adaptation with pain, providing a broader expression of pleasure and pain than usual.

At the most elementary level, a skin injury remains painful because pain receptors (e.g., free nerve endings) fail to adapt; they continue to transmit action potentials. Bright light is painful because the visual receptor does not adapt sufficiently or accommodate to the extreme stimulus in some way. Similarly, a constant, low-intensity sound is irritating because adaptation to auditory stimuli is incomplete. At a more complex level, the inability to locate a lost object is unpleasant primarily because the uncertainty of its location cannot be reduced, so the activity to find it cannot be stopped. Conversely, we derive pleasure from sensation or perception when we can reduce the sensation and conclude the adaptation activity. For example, we swallow food as a means to improve adaptation to the taste of the food; a continuous tasting experience is unsatisfying. And, of course, when we solve a problem, thereby reducing the uncertainty surrounding its solution, we derive pleasure from the act of solving it. Since the reduction of uncertainty is accompanied by a gain of information, we could say that acquiring information is pleasurable. The reduction of uncertainty represents the conclusion of the perception-action cycle, pleasurable if consummated, painful if not.[39]

---

[39] Recently, some psychologists are beginning to see that perceived entropy is closely related to anxiety and suffering (Hirsh, Mar & Peterson, 2012). (Editor's Note)

Consequences for the study of human behavior

So far, we have seen that the behavior of animals is guided by the principle of perception-action, whereby perceived uncertainty is proportionally transformed into actions aimed at reducing this uncertainty. Through our behavior, we continuously attempt to reduce perceived uncertainty, although we do not always succeed. There are three main ways to reduce perceived uncertainty:

1) Act on the environment in a way that reduces perceived uncertainty. In this case, we transfer the perceived uncertainty to the environment. This is the ideal solution for every living being (Schrödinger's theorem) since it allows the entire organism to remain informed.

2) However, 1) is not always possible due to obstacles posed by the environment in increasing its uncertainty (primarily due to the struggle for entropy in humans). When 1) is not possible, there is the option of acting on the organism itself, transferring the uncertainty to some less important organ than the nervous system. We refer to this phenomenon as "somatization."

3) When both 1) and 2) are not possible, uncertainty remains stored in the nervous system in the form of electrical charges, significantly affecting its efficiency, leading to what we call "brain disorders."

Typically, mixed solutions involving all three pathways are found. Part of the uncertainty is transferred to the environment, another part is diverted to some organ within our body, and another part remains in our brain. In fact, death occurs due to the inability to completely transfer uncertainty to the environment, meaning the progressive accumulation over the years of entropy in our body (somatization) and uncertainty in our brain (brain disorders).

PART 2. Biopsychology

Affection, health and well-being

"Nothing is easier than to admit in words the truth of the universal struggle for life, or more difficult at least I have found it so than constantly to bear this conclusion in mind. Yet unless it be thoroughly engrained in the mind, I am convinced that the whole economy of nature, with every fact on distribution, rarity, abundance, extinction, and variation, will be dimly seen or quite misunderstood."

(Darwin, 1859)

One of the most important results of my research has been the discovery of a close and direct relationship between social relationships, health, and human well-being. In this second part, I will attempt to show you how the ecology of our social relationships is directly related to non-communicable diseases and other disorders such as suicides, traffic accidents, depression, and other dysfunctions.

Ecology is the study of the relationships of all living beings with their environment, especially with other living beings with whom they share their habitat. For modern human beings, their primary environment is composed of their fellow humans, the other people with whom each one shares their life. What I have discovered is that parasitism has a significant impact on human social relationships and is the backdrop for many diseases, such as non-communicable diseases and other disorders.

While I present the biological link that connects behavior with non-communicable diseases here, I do not intend to present a new theory. In fact, this theory does not oppose findings in many fields. On the contrary, this theory provides a coherent unification of a large number of scattered facts.

This theory has three main characteristics:

1) It only involves physical and biological principles and does not mention any entity that is not material. I will not introduce any 'ad hoc' hypotheses, no 'psychological' theories, or any other type of metaphysical speculation that has not been demonstrated and accepted by the entire scientific community. Of course, this does not mean that this theory has been proven.

2) It is empirically testable. Anyone can check this theory after understanding it to try to prove or refute it themselves. I will not speak philosophically; I will not use concepts that cannot be observed and measured.

3) It is a medical theory because its main result is precise instructions for preventing many non-communicable diseases and improving our health.

I am aware of the reader's skepticism towards these words. But due to the great importance of the topics discussed, I encourage everyone, especially doctors and other healthcare professionals, to take this work seriously and try to logically and empirically refute it. Personally, I have empirically confirmed this theory for many years, and I trust in its solidity, but I would be the first to accept any empirical evidence of its weakness.

Like everything related to diseases, this theory is neither pleasant nor beautiful (beauty is not a concern of science); on the contrary, this theory deals with some cruel and repulsive behaviors, and even though it has proven effective, I do not like it at all. Therefore, I am particularly willing to accept any evidence that allows me to reject it. One of the wonderful things about science is that we only need a single fact to reject a theory, whereas any amount of facts is not enough to definitively prove it (Lakatos, 1970; Popper, 1959). The reader is welcome to send me any fact they believe could refute this theory. However, if over time this theory is not empirically rejected, I believe that a significant change in our health and well-being will be achieved in the near future.

The distinctive characteristics of the human species [40]

It is undeniable that the human species possesses special characteristics that distinguish it greatly from other animal species. However, each species has unique and distinctive characteristics. For example, elephants have a splendid trunk that no other species in nature possesses. Felines have extraordinary agility. Snakes have limbless vertebrate bodies, and butterflies have splendidly colored wings with astonishing patterns. In the case of the human species, these characteristics seem so special that they have long led us to believe that we belong to a different world, that our species does not fully belong to the natural world.

The list of special characteristics of human beings is impressively long: intelligence, creativity, technology, centered bipedalism, versatile hands, hairless and highly sweating skin, great facial expression, delayed offspring maturation, neoteny, language, culture, social organization, reasoning, written language, religion, ethics, morals, values, cooperation, compassion, art, science, affectivity, learning, monetary economics, and more.

No other species in natural history exhibits this array of special and unique characteristics. People are bewildered by such an astonishing set of features, and that is why it is of paramount importance to clearly identify how these characteristics are related to each other, what causes others, which are fundamental, and which are secondary, in order to understand human behavior. To achieve this, it is necessary to understand how and why they emerged during human evolution and their function in the species' survival.

The following diagram illustrates how the major distinctive characteristics of the human species are related and how they evolved, especially concerning the emergence of culture.

---

[40] The texts in this unit were written in the year 2007 and are part of the document "Scientific Foundations of Human Behavior Part II." (Editor's Note)

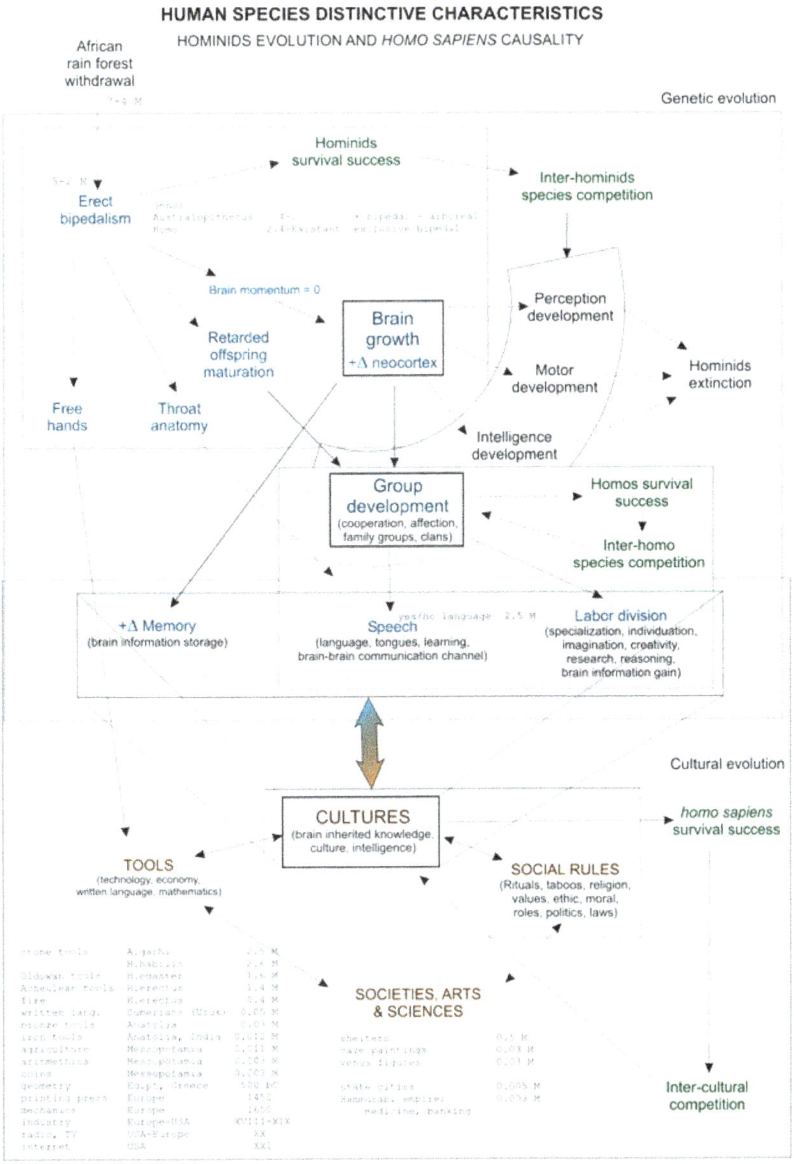

HOMINIDS EVOLUTION AND *HOMO SAPIENS* CAUSALITY

African
rain forest
withdrawal

Genetic evolution

Hominids
survival success

Inter-hominids
species competition

Erect
bipedalism

Brain momentum = 0

Retarded
offspring
maturation

**Brain
growth**
+Δ neocortex

Perception
development

Motor
development

Hominids
extinction

Free
hands

Throat
anatomy

Intelligence
development

**Group
development**
(cooperation, affection,
family groups, clans)

Homos survival
success

Inter-homo
species competition

+Δ Memory
(brain information storage)

Speech
(language, tongues, learning,
brain-brain communication channel)

Labor division
(specialization, individuation,
imagination, creativity,
research, reasoning,
brain information gain)

Cultural evolution

CULTURES
(brain inherited knowledge,
culture, intelligence)

homo sapiens
survival success

TOOLS
(technology, economy,
written language, mathematics)

SOCIAL RULES
(Rituals, taboos, religion,
values, ethic, moral,
roles, politics, laws)

SOCIETIES, ARTS
& SCIENCES

Inter-cultural
competition

Figure 6. Graphical representation of the various genetic and cultural characteristics of the human species and their evolution over the course of 5 million years of hominid evolution.

149

To begin our analysis of human evolution, we need to go back to the emergence of the first hominids. Hominid species evolved from a primate that shared a common ancestor with today's chimpanzees. Just about 5 million years ago, the recession of the African rainforest forced primates to adapt to life on the savannah, giving rise to the first hominids (Leakey, 1994; Johanson & Edey, 1990). These species were characterized by centered bipedalism, a special form of bipedalism where the body's center of gravity is vertically aligned with both feet.

Centered bipedalism provided significant advantages to early hominids, allowing them to significantly expand their field of vision on the savannah. In the past, they could seek refuge in tree canopies, but since trees were scarce on the savannah, centered bipedalism enhanced their ability to detect predators and increase their chances of moving to safer positions.

Additionally, centered bipedalism presented new anatomical opportunities for hominids, such as brain growth, the development of verbal language, and the freedom of their hands (Lovejoy, 1988; McHenry, 1994; Leakey & Walker, 1997). Among these opportunities, brain growth proved to be especially crucial because, without it, the development of language and tool-making, even if anatomically possible, would not have occurred. It's important to note that, for example, although most dinosaurs had free hands, they couldn't develop tool-making due to the relatively small size of their brains.

The growth of the brain in hominids was a result of centered bipedalism, meaning that the body's center of gravity was vertically aligned with both feet. This meant that the hominid brain didn't have to support any moments of gravitational instability.[41]

All animals, except for humans and penguins, have their brains located outside their center of gravity. This design imposes some limits on brain growth. Terrestrial animals must expend energy to support their brain proportionally to the product of brain mass and the distance of the brain from their center of gravity. On the other hand,

---

[41] A moment occurs when a force acts at a certain distance, and its value is the product of the force by its distance.

marine animals must expend energy to counteract the friction between their brain, which is always at the front, and the water. In both cases, if the brain grows, they must increase their energy consumption, which is crucial for their adaptation and survival. Considering the economic principle of nature, we can state that all animals have the minimum brain they need to adapt and survive. Unnecessary brain mass increase is always dangerous because it would entail unnecessary energy consumption.

Unlike penguins, who face water resistance, centered bipedalism in hominids allowed the brain to grow without a significant increase in energy consumption. You can recall how women from different tribes carry heavy loads on their heads because the absence of moment (distance x force) reduces the effort required to transport large masses. Therefore, by becoming centered bipeds, hominids were able to increase the size of their brains without any selective pressure forcing them to develop greater musculature to support it.[42]

In other words, the brain grew because it could grow for free. With a negligible increase in energy consumption, hominids could enjoy a significant increase in brain capacity. In conclusion, centered bipedalism offered them a splendid gift as it allowed them to develop a larger brain, and with brain growth, new skills could be acquired, enhancing their chances of survival.[43]

Recent discoveries in human paleontology indicate that there were a large number of different human species during the hominization process, which became extinct due to natural selection processes (Leakey, 1994; Johanson, & Edey, 1990). In fact, considering that the evolution of hominids spanned approximately 5 million years and offered a wide range of anatomical possibilities, it is reasonable to expect the existence of hundreds of different species. However, the

---

[42] You must bear in mind that the brain plays a central role in the survival of animals, so it will be the first to grow whenever there is an opportunity for it to do so.

[43] Natural history shows that whenever a species acquires new skills, competition among species and within them, as well as pressure from predators, forces them to evolve towards the full development of these skills. In the case of hominids, this process marked the beginning of a new evolutionary course.

low population density of hominids during this period makes it challenging to recover fossils. Nevertheless, we know that brain growth yielded a wide range of different outcomes, and it is undeniable that various hominid species evolved by harnessing their new cerebral capacities. Some became more sensitive, others more agile, some began to manufacture primitive tools, others developed more complex relationships, and others combined various skills.[44]

The question that arises is, which of these different evolutionary paths was selected by natural selection to result in the human species? Which abilities significantly improved the survival of hominids?

Let me briefly discuss each of them. First, improved sensory processing enhances the survival of any species by allowing for the quicker detection of predators and sources of food. Likewise, increased agility in movement helps in escaping predators and reaching challenging places or hunting faster animals. However, modern Homo sapiens sapiens is not characterized by exceptional sensory processing or agility, and in fact, prehistoric hominids were anatomically far from developing such skills to successfully escape the attack of a group of lions, for example.

On the other hand, it is evident that the development of intelligence is useful for survival in any case. However, it cannot be achieved solely through brain growth but also requires the ability to learn, and learning necessarily involves making mistakes. Unfortunately, in nature, errors often have fatal consequences, and there is no opportunity to improve intelligence. For this reason, most animal species do not possess a high level of intelligence as they do not have the chance to learn from their mistakes.

However, it is doubtful that an intelligent hominid, alone and armed with rudimentary tools, could face the attack of a predator or find enough food by itself. The survival chances of a solitary individual in a hostile environment like the African savannah are truly

---

[44] The brain is a computational machine and can be dedicated to enhancing various skills such as sensory processing, agility in movements, intelligence, and social organization.

slim, which is why social skills were a key factor in the evolution of hominids (Boehm, 1999; Dunbar, 1996). Social organization maximizes the odds of survival because the strength of a well-organized group far exceeds any individual advantage. Predators have a harder time attacking an organized group, and the chances of finding food are also greater.

Thanks to the new anatomy of centered bipedalism, not only could the brain grow, but also oral language could emerge as a new and powerful channel of communication in hominid groups. This was because to significantly improve social organization, a powerful communication tool was necessary. Furthermore, brain growth allowed for increased memory storage, which, coupled with the development of language, enabled the transfer of information between hominid brains.[45]

In conclusion, the new mass of neurons could have been dedicated to improving perception skills, mobility, manufacturing, reasoning, and more. However, natural selection favored individuals with better social skills through language (Falk, 2004).[46] Among different hominid species, only those with brains more capable of learning and developing linguistic skills prevailed. Before hominid species with great manufacturing abilities (let's say better weapons) prevailed, other species with stronger social organization (more cooperation) succeeded.[47]

Thus, we understand that the most important selective pressure among different hominid species was social ability. Only strong societies conquer the weak. The collapse of the USSR was not due to a lack of nuclear weapons but rather the low level of its social

---

[45] Centered bipedalism and brain growth focused on improving social skills led to a surprising phenomenon: the emergence of culture as an independent form of life. Although culture is a phenomenon present in some other species (Sabater, 1978), human cultural development is unparalleled in the history of life on Earth.

[46] Language plays a fundamental role in social organization, and the human brain acquires basic linguistic skills only after its first year and a half of life.

[47] You can think of the Vietnam War as an example. The United States lost that war because its society did not cooperate in the war in the same way that the Chinese and North Vietnamese societies did. Social cohesion is the most powerful weapon.

organization, especially its economic organization. In contrast, China is improving its economic organization to avoid the same fate. The success of the US lies in its ability to integrate people from very diverse cultures into a large society with a common identity. Conversely, the weakness of the EU is due to its inability to build a new large society because it cannot give up its different identities in favor of a common and new one. The Catholic Church was a great power due to its strong social organization until modern states with stronger social organization emerged. "Unity is strength" is the motto that defines the human species.

Let's think about a simple pencil. How many people have had to intervene for us to have a pencil in our hands? Hundreds, thousands of people have had to intervene. And it's the same for everything we consume to enjoy our well-being.

In conclusion, it is undeniable that the new neural capabilities allowed by brain growth were dedicated, through natural selection, to building complex societies. In other words, centered bipedalism is responsible for our great social skills, and these are truly the most distinctive feature of the human species. We could think of Homo sapiens sapiens as the species whose brain is genetically better prepared to develop social skills.[48]

The evolution of the human species has been driven by the construction of increasingly complex societies, from small tribes formed by a few family groups, through the first villages, cities, and states, to large states and communities of states. The immense human societies of today form the basis for current cultural and economic development.

It is not the individual but society that is responsible for all human achievements. Knowledge, language, intelligence, technology, the arts, religion, politics, or science are the result of our social life and not our individual abilities. A human individual is a great artist or a great

---

[48] It is likely that our brain has genetically preconfigured preneuronal circuits to facilitate the development of language and cooperation (Pinker, 1994; Hauser, Chomsky & Fitch, 2002).

154

scientist not because they have special genes but because they have received the skills accumulated over generations from their fellow human beings. Each human individual is the product of the entire history of their society. This is the most important characteristic that distinguishes the human species from others.

The social nature

As we have seen, what characterizes us as human beings is not intelligence or reasoning, but rather the social skills that allow us to establish intricate networks of mutually beneficial relationships in increasingly complex societies.[49]

But what does it mean to be social? Being social means that individuals are not capable of surviving alone and need the support of others to survive and reproduce (Dunbar, 1998; Trivers, 1971; Wilson, 1975). In the course of evolution, many species have evolved towards different forms of social organization. The evolution from solitary living to group or community living is very common in nature. All multicellular living organisms result from a complex community of cells that in early stages of life lived alone (Margulis, 1970). There are a large number of social species with very different degrees of social need and organization. Many species are only social during part of their life (usually while they are young) and then become solitary individuals. For example, bears are a social species only during the early years of life when the cub needs the mother's help to survive. When the mother abandons it, the bear will live in complete solitude, except for the inevitable encounters with other bears, which are usually more or less aggressive.

Other species are social throughout their entire lives. Species like ants, lions, or wolves are highly social because they cannot survive without the collaboration and assistance of other individuals of their species. Of course, the degree of complexity and social need varies greatly from one species to another. Among mammals, humans are

---

[49] Mutualistic relationships are interactions in which A and B cooperate for a common goal, whether it is to obtain a resource or avoid a predator. In this case, it is essential that an entity C, which is not in a mutualistic relationship with A and B, is harmed, as otherwise, the second law of thermodynamics would be violated. There is a temptation to confuse mutualism as a form of non-harmful interaction and, therefore, ideal. But nothing could be further from the truth. Mutualism is a strategy that provides significant benefits to the cooperators because it allows them to more easily obtain the necessary resources at the expense of other organisms not belonging to the cooperating group.

undoubtedly the most social species. This means that a human cannot survive alone without the direct and indirect collaboration of other individuals of the same species. From birth, humans constantly need the collaboration of their fellow beings. Of course, this social dependence has its benefits because, through collaboration, the group becomes stronger, and the individual has a better chance of surviving and reproducing.

But what is the cause of living in groups? The answer is simple: to avoid extinction. If a species of solitary individuals can survive without major problems, there is no reason to develop social organization. But if their survival is threatened, then a good alternative might be to develop some form of social organization because, by living in it, any individual could receive help from their peers to survive. It is under this pressure that living beings adopt a cooperation interaction with each other (Dawkins, 1976; Wilson, 1975). The result of this cooperation is mutual benefit obtained from the losses of third-party organisms. The success of the group, and therefore of the human species, lies in the exchange of support among its members.

In this sense, we can say that all social organizations are based on the exchange of support among their members in order to achieve survival (Kropotkin, 1902). Receiving support from others increases the chances of survival and is a good strategy if one cannot survive alone. A concrete example of this can be found in the social organization of lionesses. A group of lionesses exchanges support among themselves for hunting, raising their cubs, and defending their hunting territory, among other benefits. Living alone, each lioness trying to hunt, raise her litter, and defend her territory could not survive and reproduce successfully. If they lived alone, lions would go extinct. For the same reason, the human species lives in groups and societies because none of us could survive if we lived isolated without the support of others.

Building upon everything mentioned earlier, we can define social organizations as systems of support exchange among their members, regardless of the particular form they use to achieve it (Kropotkin,

1902; Emerson, 1976). In this context, we can also define social support as any action performed by an individual that increases the chances of survival for one or many other individuals who are the recipients of the support (House, Landis & Umberson, 1988; Cohen & Wills, 1985).

This means that we recognize the existence of a social organization only if we can observe the presence of some form of support exchange among its members (Barrera, 2000; Durkheim, 1893; Putnam, 2000). A social organization is not just any group of individuals interacting with each other in the same place. We need to observe some form of support exchange among them to enhance their chances of survival. Of course, within any social organization, there is competition and predatory interactions among its members, but what makes them social is the presence of some degree of support exchange. For example, if we are observing a group of people fighting on the street, we cannot say that we are observing a social system. We could say that we are observing two social systems fighting against each other if we could observe that a subset of people is supporting each other to fight against the other subset, which is supporting each other to fight against the first. In this sense, we would say that there are as many social systems as there are groups of people supporting each other. In summary, the social nature is characterized by the exchange of support.

The nature of social support [50]

The transfer of social support forms the basis of the human social structure (Durkheim, 1893; Godelier, 1996; Kropotkin, 1902). Humans are the species with the most advanced social development, which implies that we are the species with the greatest need for support.

Currently, no adult person could survive on their own without the help of their fellow humans, and, of course, it would be impossible for them to maintain the quality of life they enjoy in their social life. Needless to say, children, the disabled, or the elderly have even fewer chances of survival outside the social structure. In conclusion, the exchange of support is the glue that holds the human social structure together and is intrinsically linked to being human.

---

[50] The texts in this unit were written between 1998 and 2002. They are part of a series of research projects conducted with the collaboration of Pilar Gonzalez. (Editor's Note)

Affection as social support

In general, affection is often equated with emotion, but in reality, they are very different phenomena, although they are undoubtedly related to each other. While emotion is an individual subjective response that informs about the probabilities of survival offered by each situation (Wukmir, 1967), affection is a process of social interaction between two or more organisms (Panksepp, 1998).

From our everyday use of the word 'affection,' we can infer that affection is something that can be given to another. We say that we "give affection" or that we "receive affection." Thus, it seems that affection must be something that can be provided and received. In contrast, emotions are neither given nor taken away; they are only experienced within oneself. We typically describe our emotional state using expressions like "I feel tired" or "I feel great joy," while we describe affective processes as "I show affection" or "I give a lot of security." In general, we do not say "I am given emotion" or "I am given feeling," but we do say "I am given affection." Furthermore, when we use the word 'emotion' in relation to another person, we say "so-and-so excites me" or "so-and-so produces such and such emotion in me." In both cases, it primarily refers to an internal process rather than transmission. It seems that a fundamental difference between emotion and affection is that emotion is something that occurs within the organism, while affection is something that can flow and be transmitted from one person to another.

Unlike emotions, affection is something that can be stored or accumulated (Bowlby, 1969; Harlow, 1958). For example, we use the expression "recharge our batteries" during vacations to refer to the improvement in our disposition to attend to our children, friends, clients, students, colleagues, etc. This means that in certain circumstances, we store a greater capacity for affection that we can give to others. It appears that affection is a phenomenon like mass or energy, which can be stored and transferred.

On the other hand, our experience teaches us that giving affection requires effort. Caring for, helping, or understanding another person cannot be done without effort. Sometimes we may not realize it, but in most cases, we all experience the more or less intense effort we make to provide well-being to others. For example, taking care of someone who is sick requires effort and is a way of providing affection. Trying to understand someone else's problems is an effort and is another way of giving affection. Trying to please another person, respect their freedom, or make them happy with a gift are actions that require effort, and all of them are different forms of providing affection.

However, despite the differences, affection is closely linked to emotions because similar terms can be used to express an emotion or an affection. For instance, we say, "I feel very secure" (emotion) or "I receive a lot of security" (affection). It seems that we designate the affection received by the particular emotion it produces in us.

When we commonly say that human beings need affection for their well-being (Baumeister & Leary, 1995; Fromm, 1953; Maslow, 1943), we are actually referring to the fact that they need the help and cooperation of other human beings to survive. In other words, the need for social support is expressed as a need for affection or an affective need. Hence, affection has been considered something essential in the life of every human being since the origins of our culture (Aristotle, 2002; Rousseau, 1762). Giving affection means helping others, promoting their well-being, and their survival. Indeed, affection, understood as social help or cooperation, aligns with the characteristics we attribute to it in our everyday language. Social support is something given to others and involves an effort for the one providing it.

Social support as work (W)

The importance of social support in group and organizational dynamics is undeniable (Lewin, 1936; Janis, 1972; Sherif, 1966). As we defined earlier, every social group or organization is based on the exchange of support among its members to enhance their survival. There are countless ways to provide support, and it would be tedious to decipher them all. What I want to teach you now is the common process underlying all support exchange phenomena.

Our experience teaches us that providing support is something that requires effort. No one can provide support without effort. Respecting, caring for, helping, or understanding another person cannot be done without effort (Bowlby, 1969; Erikson, 1950; Maslow, 1943). Sometimes we may not realize this, especially if the relationship is sporadic or in its early stages. The excitement of a new relationship, for example, may not allow us to see the effort we make to please the other person, to provide them with well-being. However, in general, when a relationship is stable and evolves over time, we all experience the effort we must make if we want to provide affection to the other.

In conclusion, to truly help another person, some kind of work[51] must be done on their behalf, which is why we say that providing affection requires effort. Of course, there are many ways to provide affection since a person can perform very diverse tasks for the benefit of others. But ultimately, social support always involves physical work[52] or the exchange of energy. There is no way to support others

---

[51] The way living beings behave is by working in their environment. Although this change in wording may seem minor, it has significant consequences for understanding behavior. Work is the means by which systems transfer energy to other systems, and this concept opens the door to a physical and biological analysis of behavior.

[52] The concept of work has a precise definition in physics: any means by which energy is transferred from one system to another. Therefore, work has the same units as energy. The difference between energy and work is merely contextual. We speak of energy as a characteristic of a system, and we speak of work as a characteristic of any process of energy transfer.

without doing some kind of work and expending a certain amount of energy.[53]

As we saw earlier, living beings need to obtain free energy[54] from their environment to survive, and we obtain this energy through interactions with the environment that involve an exchange of energy (entropy). In the case of the human species, we can distinguish two ways of obtaining this energy:

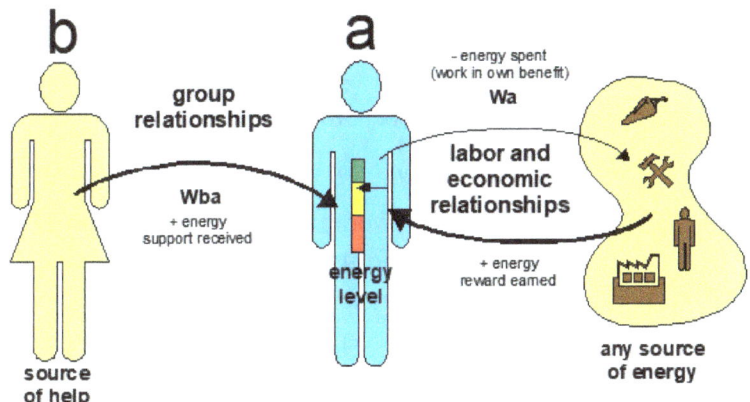

Figure 7. Two sources of energy: 1) working in our environment to obtain more energy and 2) receiving support from our peers.

1) We can obtain this energy by working on our environment so that the energy gained is greater than the energy expended. In these cases, we say that we are working for our own benefit. We represent the work done by person a for their own benefit as W_a. Labor and business relationships fall into this type of interaction. In a business relationship, for example, a merchant sells products for less money

[53] In reality, what we exchange in our human relationships is entropy, not energy. When we say "we work for the benefit of the other," we are referring to the transfer of entropy from the other person to us. This is the correct way to distinguish it from work that does not contribute to the other person's survival.

[54] Free energy is the amount of useful energy available to do work, and it is what living beings "feed" on to stay alive, that is, to stay away from illness (entropy) and death (thermodynamic equilibrium). (Editor's Note)

than they cost, in order to make a profit. The interaction between the seller and the consumer is not for the benefit of the consumer but for the benefit of the seller. Of course, the consumer can derive benefits from their purchases, but it will be after this relationship and will involve other interactions.

2) But as social beings, we can obtain part of the energy we need by receiving it from other people through our group relationships, W_{ab} and W_{ba}. This second source of energy is what we call affection or 'social support.' When we provide support, we are working for the benefit of others without receiving a direct reward. We represent the work done by person a for the benefit of person b as W_{ab}, and we represent the work done by person b for the benefit of person a as W_{ba}

All living beings have to work to survive. The energy and resources we need do not come to us spontaneously, so we have to work hard and make a great effort to obtain them. In social species, individuals can obtain some of the energy and resources through the support of their peers. This is why social organization has more advantages than living alone.

By providing support, we transfer energy to other people for their survival, working for them for their benefit. By working for the benefit of their children, parents provide the necessary energy for them to grow and mature. As children grow, they become more capable of working for their own benefit to obtain energy, and only then can parents reduce the work on their children. Similarly, in all our group relationships, we are exchanging support with our peers, that is, we are doing work for their survival, and we are receiving work from them for our survival.[55] But working always means expending or losing energy, so our lives can be compromised if we do not obtain enough energy to offset our losses. Supporting others is good and positive, but we lose energy in the process.

---

[55] Of course, not all work done on others increases their chances of survival. Here, we are only referring to work that increases the chances of survival.

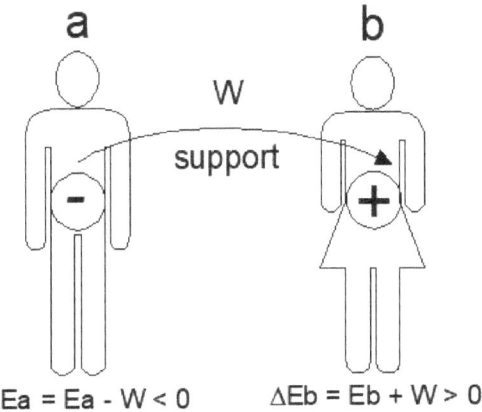

$$\Delta E_a = E_a - W < 0 \qquad \Delta E_b = E_b + W > 0$$

Figure 8. This figure shows how social support, understood as work (W), ideally results in an energy loss for the person providing support equal to the energy gained by the recipient of the support.

When providing support, that is, working for the benefit of others, the supporter loses energy while the recipient of the support gains it. This is because energy is a conserved quantity. In ideal conditions,[56] we could say that the one receiving support gains the energy the other has lost. However, ultimately, there is always an imbalance behind any social interaction—someone performs work and loses energy, while someone else gains energy thanks to the work of the former. Of course, there is an exchange of support in any relationship, so the final balance could be equalized. But what I want to demonstrate here is that each individual act of support is unbalanced because it can only be done by working for the benefit of others.

The balance and economy of these different sources of energy and the various ways of spending it are the central question for our survival, health, and well-being.

---

[56] In real conditions, we know that thermodynamically it is impossible to convert all work into useful energy since entropy is produced, and therefore, the useful energy gained is always less (in absolute terms) than the useful energy spent.

The impact of our exchange of support through our group relationships has a significant influence on our energy balance. We just need to remember that without this source of energy, we couldn't survive. And if our energy balance is systematically negative, if we expend more energy than we obtain over an extended period of time, then our health and well-being will deteriorate.

Forms of work

Human beings are the most complex phenomenon we know in the universe. Therefore, when we try to introduce the concept of work to study human behavior, we must realize that we are not talking about simple work machines like a car engine. Additionally, we do not pretend to be able to measure human work because it is very sophisticated.

In general, we can distinguish three major types of work in humans: chemical work, mechanical work, and computational work. Of all these, we are mainly interested in the third one, which is cognitive or brain work.

Chemical work is performed by all the metabolic processes in our body. It is responsible for maintaining our hardware in good condition. We expend a lot of energy on this type of work and need to continuously obtain energy to carry it out. Our body, including the brain, is continuously losing its structure and needs to be repaired at all times.

Mechanical work is all the work done by our muscles. Through them, we can move masses at a distance both inside our body and outside of it. Before the Industrial Revolution, humans expended a great deal of energy on mechanical work, and it played a very important role in health and well-being. However, after the widespread use of machines that replaced our muscular work, we now expend very little energy on it. Paradoxically, modern humans suffer health problems due to a sedentary lifestyle and the underuse of their muscles. Therefore, in the study of modern human behavior and its relationship with health, this type of work is generally not significant, although it may be in some special cases.

Finally, we have to consider computational work, that is, the work performed by our nervous system, especially our brain. This work involves making decisions and directing all the chemical and mechanical work to maintain the health of the organism.

As we have seen before, brain work involves obtaining information. You might find it difficult to understand brain work in terms of energy, but it will be easier to understand in terms of information. Information could be seen as another form of energy, such as kinetic, potential, chemical, electrical, etc. In short, we could say that the brain functions by transferring information to other systems and, to do so, it needs to obtain information from its environment.

From another perspective, we can observe that our brain has a lot of tasks to perform, which can be classified into three sets. One set is responsible for controlling the internal state of the organism, i.e., controlling the chemical work of the metabolism of our body. To perform these tasks, the brain mainly uses hormonal glands. The second set of tasks is concerned with controlling the mechanical work of the muscles to obtain chemical energy from the environment. That is, obtaining the energy needed to carry out all the chemical or metabolic work. Finally, the brain has to work to obtain the information it needs to function properly and perform all its tasks.

Now let's go back to the concept of affection and social support. Remember that earlier we said that giving affection means performing some kind of unpaid work for the benefit of another person. When an organism does work, it consumes a portion of its available free energy, proportional to the magnitude of the work performed and the efficiency with which it is done. Well, this work not only consumes metabolic energy but also consumes part of the brain's processing capacity. This means that most tasks require using the brain to coordinate all the actions involved in performing the task.

The affective capacity of humans

The affective capacity of each individual is determined by their ability to work for the benefit of others without compensation. An individual's ability to help others is limited because it directly depends on the amount of resources they have access to and their capacity to perform work. Therefore, we can say that affective capacity (social support) is something that can accumulate and can vary over time and from one individual to another, as both available resources and the capacity for work are cumulative variables[57] (Berkman, et al., 2000).

Regarding brain work, there are significant differences based on age, as computational capacity is acquired through learning (Piaget, 1950). The computational capacity of a newborn is very low and gradually increases through their learning experiences.

Furthermore, there are significant differences among adults depending on the degree of development they have undergone, i.e., the amount of information stored in their brain. Education, learning, training, or studies determine an individual's capacity and ability to solve problems (Vygotsky, 1978). So in any society, we find individuals with very low levels of education and a few (generally) with a high level of cultural development. Of course, I am referring to development in the broadest possible sense, not just academic knowledge or economic skills, but the ability to solve existential problems (survival and quality of life) of any kind.

Finally, since affection is the general way in which we express the need for social assistance, affection needs vary from one individual to another (Bowlby, 1969; Cohen & Wills, 1985). Thus, socially more dependent individuals, such as children, the elderly, or the sick, are the groups that need more affection to survive. On the contrary, adult individuals who have undergone appropriate maturation require much less affection and, consequently, can provide more affection to others.

---

[57] If emotion behaves like an intensive state variable, affect behaves like an extensive state variable (the total value equals the sum of its parts).

As we will see later, providing support is not easy at all, and in many cases, the willingness to provide it is much greater than the actual capacity to do so. This is the main reason why most of us have developed with more or fewer gaps due to the lack of adequate social support and, consequently, continue to need it despite being adults.

Social support as a biological resource

In biology, we call anything that living beings use to survive a resource. All living beings require specific components or elements (forms of matter and energy) to stay alive. The absence of a particular resource can endanger the life form that depends on it and even lead to death if the situation persists.

Due to the existence of an infinite diversity of life forms, there is also an infinite diversity of resources. Each species or life form specializes in particular types of resources and specific ways of utilizing these resources, which ecologists refer to as the ecological niche of the species (Hutchinson, 1957). For something to be considered a resource, it must be a physical and limited element, meaning it's a quantity of something that can decrease or increase and thus be more or less available to the living beings that use it. When a living being uses a resource, it reduces the quantity available to other living beings that depend on it. Additionally, a resource is something that is consumed by living beings. Consumption doesn't necessarily mean ingestion but degradation in the process of using it to sustain life. For example, the process of feeding results in the degradation of the resources ingested, just as in respiration, the use of territory, the use of defenses, or the use of tools. All resources degrade when used by living beings (Odum, 1971; Tilman 1982).

Now, let's see if we can consider the hypothesis that affection is a biological resource. It's not about proving it but rather seeing if it makes any sense. To do this, let's examine whether affection can satisfy the properties that define a biological resource.

Could we survive without affection? I believe everyone agrees that it's not possible to survive without the affection of others, or at least I don't know anyone who thinks otherwise. Even if we have enough food, shelter, territory, knowledge, or money, experience teaches us that it's not possible to maintain a sufficient state of health or a satisfactory life if we don't also have the respect, affection, attention, understanding, friendship, love, help, or protection of some people around us. In fact, life becomes unbearable without the affection of others, and clinical cases are full of examples demonstrating the crucial importance of emotional relationships for human well-being.

If we divide living beings into social and asocial, meaning those that require the cooperation or participation of other individuals of their species to survive (social) and those that do not (asocial), then we can say that social living beings need a resource that asocial ones do not have, namely, the affection or cooperation of other individuals of their own species.

There is no doubt that affection meets one of the conditions to be considered a biological resource: survival is impossible without it. However, we need to determine if affection can also be something limited, physical, a quantity that can decrease or increase, and if its use implies its consumption and degradation. If affection also meets these conditions, then there will be no obstacle to considering the hypothesis that affection is another biological resource.

Until the 20th century, there wasn't enough evidence to understand that affection is also a physical process. It was believed that affection was something spiritual in the sense that it was not constrained by matter or physical laws. However, throughout the 20th century, Neurobiology has provided a wealth of scientific evidence that all our behavior and subjective experience (feelings, emotions, thoughts, etc.) are exclusively physical phenomena[58] that occur in our nervous system (LeDoux, 1996; Damasio, 1994; Panksepp, 1998).

Therefore, when a person provides us with affection (joy, zest for life, understanding, etc.), both they and we experience physical transformations in our nervous system and throughout our entire organism (Damasio, 1994). It's not an event that occurs independently of our bodies; it's a physical event that takes place within our bodies. Fundamentally, the physical processes intimately involved in our emotional experiences (both positive and negative) are highly intricate

---

[58] Emotions are the result or manifestation of complex physical processes in our brain, as we have seen before. Therefore, we can easily alter our emotions through a multitude of drugs that chemically act on our brain.

electrical flows that traverse our entire nervous system (LeDoux, 1996). The primary form of affection that we know of is thus an electrochemical phenomenon that occurs in the brains of all animals (Panksepp, 1998). Affection is physically obtained through the senses (perception), physically transferred through behavior electrically governed by the brain, and chemically and physiologically stored in the brain through neural circuits.

Giving or receiving love, affection, attention, care, understanding, support, or respect are manifestations of positive affection that directly affect the electrochemical state of our brains (Fredrickson, 2001). This means that if I feel happier today than I did yesterday, it's because the electrical processes occurring in my nervous system are very different from those that occurred yesterday. My joy or sadness are expressions or manifestations of different physical processes within my body, particularly in my nervous system. Nowadays, no one doubts that our emotional experiences are the result of exclusively physical processes, and therefore, affection, which is a process that alters our emotions, is also a physical phenomenon.

In summary, we know that affection is an exclusively physical process, even though we don't understand it in detail or can precisely map out this process. But is affection limited? Does affection deplete? Does affection degrade? Since we don't have any instrument to measure quantities of affection, we have to rely on our experience to answer these questions. For example, if affection were inexhaustible or unlimited, there would be no scarcity of affection in humans; everyone would have large quantities of affection, joy, zest for life, love, and friendship. We would be able to provide significant amounts of affection to others without any detriment to ourselves, etc. Unfortunately, experience teaches us otherwise. If we observe our own lives and those of the people around us, we must conclude that affection is something very scarce and hard to come by. Therefore, we say that affection is limited if we are to heed our knowledge and experience.

It will also be easy for us to understand that affection is something that degrades. Firstly, our experience teaches us that negative affection exists, such as contempt, insults, mistreatment, aggression, marginalization, disregard, etc. If we were to search for candidates for degraded affection, we wouldn't find anything more suitable. However, degraded affection must result from the consumption of positive affection for us to truly consider affection as a biological resource. This means that the people who insult, scorn, aggress, marginalize, or mistreat the most are likely the ones who receive the most love, consideration, affection, respect, protection, or help from others, probably even from their own victims.

On this point, experience provides us with two distinct and seemingly contradictory cases. On the one hand, we all know that the more someone is loved, pampered, or attended to, the more uncompromising, demanding, and less kind they may become. The most extreme cases can be found in great dictators, individuals who have received enormous amounts of support, respect, and affection from a large population, yet their actions have caused significant harm and suffering proportional to the affection they received. However, we don't need to go to such extremes. If we take the time to carefully observe someone in our vicinity who tends to be demanding, authoritarian, dissatisfied, or unpleasant toward us, we will soon realize that it is often ourselves who provide them with constant affection, respect, acceptance, or love that they may not truly deserve.

On the other hand, experience also shows us that not all aggressive and unpleasant individuals have received a lot of affection; quite the opposite, many of them have received a lot of negative or degraded affection, insults, contempt, mistreatment, etc. The problem lies in the fact that degraded affection circulates and accumulates in certain parts of society (marginalized sectors) through various emotional relationships, much like industrial waste circulates and accumulates in specific areas of the territory (landfills).

Thus, the expression of negative affection by a person does not necessarily have to be because they have consumed affectionate resources but simply because they have received degraded affection produced by other people who may not have a direct relationship with the affected person. In fact, most explosive manifestations of degraded affection (violence, aggression, mistreatment, etc.) result from its progressive accumulation through emotional relationships. We could say that in society, there are significant "pockets of emotional waste," where a large number of people accumulate the emotional waste produced in society, and they can hardly survive with dignity.

The same applies to industrial waste. A significant portion of hazardous waste from Western societies is deposited in countries with low industrial capacity. If we were to visit these countries, it would not be correct to assume that they are heavy energy consumers simply because they have a lot of industrial waste. It's just that the waste is not theirs; they have had to accept it due to their vulnerability compared to much stronger producers. Therefore, the facts do not seem to contradict the idea that affection degrades as a result of its consumption; quite the opposite.

In summary, we see how affection can meet all the requirements to be considered an essential biological resource for the survival of social animals, especially for humans (House, Landis, & Umberson, 1988). In fact, I believe that social support is the most important biological resource for the existence of human beings and, in general, for all social species.[59] The significance of this resource lies in the fact that individuals in social species are not born with sufficient development to fend for themselves and they need the support of other individuals in their group for a certain period until they reach adequate development. The importance of this resource in humans is evident from the number of words we use to refer to it depending on the context of the relationship. We call it "love" in romantic and parent-child relationships, we refer to it as "affection" in family

---

[59] There is a growing body of research suggesting that social support and affection are important for physical and emotional well-being. These studies support the idea that they can be considered essential biological resources for survival and well-being (Cacioppo & Hawkley, 2009; Holt-Lunstad, Smith, & Layton, 2010). (Editor's Note)

relationships, and in friendships, we use terms like "solidarity," "care," "help," "affection," "understanding," "support," "cooperation," "assistance," and more.

Negative affection

Given that we have identified affection as free energy that we receive or give to other people, we can state that negative or degraded affection (insults, mistreatment, contempt, disdain, pessimism, etc.) is a manifestation of entropy (disorder). Therefore, we have the need to transfer it to the people around us. Let's not forget that when we talk about affection, we are actually referring to physical processes that occur in the brain. The popular saying that it's not good to keep one's sorrows to oneself directly relates to the dissipative function.[60]

To the extent that we are able to transfer to others the negative affection: 1) that we produce as a result of our consumption of affection and 2) that we absorb from others as a result of our affective relationships, our state of affective health will be good. The problem is that others have no interest whatsoever in our negative affection (entropy) and therefore try to prevent it. The result is an underlying struggle among humans for affection, which we will address throughout this work.

At this point, you may already have an inkling that illness is intimately related to the systematic increase of negative affection (entropy) stored in the brain. If a person does not dissipate their negative affection and absorbs that of others, they do nothing but systematically increase their cerebral entropy. Given the crucial importance of the brain in regulating all the functions of the body, this situation ultimately leads to illness and death. In fact, people who die or become ill at a young age are usually extraordinarily kind and positive towards those around them. They rarely emit negative affection towards others and tend to be good receivers of the negative affection of others. On the contrary, people who become ill and die very late in life are often very authoritarian, dogmatic, and uncompromising, if not violent. In short, people who have no

---

[60] The dissipative function, or overcompensation function, is the function that living beings perform to dissipate their entropy into the environment and thus maintain a state of extremely low relative entropy. (Editor's Note)

difficulty dissipating their negative affection. In fact, we will see how the persistent lack of positive affection, or in other words, the failure to dissipate negative affection, leads to illness and death in humans.

The exchange of social support [61]

To understand the nature of social support exchange, we must remember that it is a specific manifestation of Schrödinger's theorem, which means the need to transfer entropy to the environment to keep the organism alive. In the case of social species, especially in humans, the environment consists primarily of members of the same group, organization, community, or society. Therefore, the transfer of entropy to the environment materializes in social relationships, with the exchange of social support being the primary vehicle for this transfer.

Thus, providing social support involves absorbing entropy from the individual receiving support, thereby helping them survive or improve their quality of life. It is essential to understand that the benefit gained by the individual receiving social support is always less (in absolute terms) than the harm incurred by the one providing it. This result is a direct consequence of Schrödinger's theorem, and although it may seem surprising, readers can empirically verify it by setting aside social biases on the matter.

It is a common belief to think that providing social support either carries no harm or even benefits those who provide it. This belief stems from the crucial importance of social support in humans, and its function has been to encourage its provision to the fullest extent while hiding the harm it entails. However, in today's society, our social development allows us to become aware of the processes involved in providing social support without jeopardizing its prevalence in our societies.

---

[61] The texts in this unit are a compilation of documents that were written between 1998 and 2003. (Editor's Note)

The struggle for social support

If we start from the hypothesis that affection is a biological resource, which is physical, limited, consumed, depleted, and cannot be generated out of thin air, this necessarily implies that affection is a phenomenon governed by the laws of nature. In essence, our emotional relationships are governed by the fundamental principle of biology, which is the principle of the struggle for life. Applied to our emotional relationships, it becomes the principle of the struggle for affection.

Survival and well-being depend directly on receiving sufficient support from our peers through group relationships (Berkman & Syme, 1979; House, Landis, & Umberson, 1988). In childhood, support is needed to reach an appropriate level of maturity. In old age, support is needed to survive (Bowlby, 1969; Seeman, 2000). Also, in mature years, we need it to compensate for any deficits in our development. When we are sick or facing any kind of problem, we need support to overcome them (Cohen & Wills, 1985). At any point in our lives, we need some form of support from our peers. At times, we may need a lot, while at others, we only need a little.

From the fact that we always need support for survival and well-being and that providing support involves a expenditure of effort and energy, we can clearly deduce that there is a continuous struggle to obtain it.

Since in modern societies, support is mainly provided through mental work, this struggle is highly sophisticated and covert, not easy to identify. We don't need to kill or act violently to obtain support. The way we compel others to support us is more refined. We can seduce and promise significant rewards to receive real support from others (Goffman, 1959). Or we can appear very weak and helpless, generating sympathy for ourselves (victimizing) and forcing others to support us (Taylor et al., 2000). We can emotionally blackmail to get support (Forward & Frazier, 1997). Or we can simply wait for others to solve the problem. There are hundreds of ways or strategies we use to coerce others into giving us the support we need to survive. In general, all these paths do not involve the use of physical violence. In most cases, we use strategy, cunning, and seduction. However, there is still a significant residue of physical violence in some situations, mainly in family relationships. Child abuse and abuse of women are widespread in our modern societies (Finkelhor & Ormrod, 2001; Heise et al., 1999). That is, although it is not the most common behavior, one can force others to support them using physical violence.

The struggle for social support begins right at birth when the baby cries for the mother to provide milk and the necessary shelter to sleep. From this moment on, the child will have to find the most efficient way to obtain social support from their parents or caregivers, and most of the time, they will have to compete with other siblings and family members to get the support they need. This is the most prototypical example of emotional competition, as siblings share the same source of affection (the parents). Among siblings, there is strong competition to receive as much affection as possible from the parents, and significant inequalities between them often arise due to this interaction (Brody, 1998). However, both end up losing because if they were alone, they would receive larger amounts of affection from their parents.

But emotional competition is not limited to the interaction between siblings. Any two individuals who receive affection from a third person engage in a competition for affection between themselves. Married couples compete for the exchange of support, children and household responsibilities, grandparents compete with sons-in-law and daughters-in-law for the support of their grandchildren, and so on.

The reason why there are constant crises and conflicts in group relationships is the struggle for the exchange of social support. Our social life is filled with deception, crises, breakups, tensions, conflicts, and separations because we are continually immersed in this struggle for the affection and social support of our peers.

To achieve success in the struggle for life, living beings adopt a wide variety of strategies. Among them, the most important is camouflage. Through camouflage, a living being can approach and obtain resources from another while evading the victim's perceptual defenses. It is a general strategy, and there is no species that does not use it to some extent. In humans, camouflage is widespread, and deception at all levels is one of the keys to success in obtaining emotional resources.

For example, the victim attitude encourages a positive response from the social environment. Many people reach old age having adopted this attitude throughout their lives. Those around them feel obligated to endure, help, sympathize with, and support those who present themselves as victims. The result is that the supposed victim survives, while the people who supported them become ill and die at a younger age. At other times, we act as if we are important, powerful, and generous, and our interlocutor cannot help but give us affection in order to gain greater benefits. We also resort to violence, to the authority we have over someone, to demand that they provide us with the affection we need. Sometimes, we give large amounts of false affection that our interlocutor cannot detect, and then they cannot avoid giving us affection to avoid feeling guilty. In short, there are many ways to make another living being, another person, provide us with affection. And all the affection that is given is lost, it ceases to be available to the giver and becomes part of the receiver.

As in any struggle, there is always a winner and a loser, and the survival of some individuals is based on the illness and death of many others.[62]

This situation is particularly critical in humans due to the current isolation from other species. Before the Industrial Revolution, when the population was predominantly agricultural, people could derive affection from domestic animals and the natural environment (plants, insects, etc.). They could discharge their aggressiveness (emotional

---

[62] Schrödinger's theorem states that for living beings to exist away from thermodynamic equilibrium, they must generate more entropy in their environment than they manage to reduce within themselves. This idea is the fundamental concept of life, life is not free, someone in our environment must pay the price for our existence. (Editor's Note)

waste) onto animals and plants. But now, the vast majority of human beings in industrial societies do not interact with other species, and therefore, they must extract the affection they need from their fellow humans. This creates a special tension in emotional relationships (mainly within families) with significant consequences for human survival, as we will see next.

Our social relationships with our fellow humans are characterized by the struggle for affection. We all try to obtain help, attention, understanding, respect, affection, support, care, or love from the people around us. We feel good when we get it, when someone devotes themselves to us, solves a problem for us, listens to us attentively, encourages us, in short, provides us with affection. We feel bad when no one pays attention to us, understands us, supports us, encourages us, or when someone pressures us, reproaches us, or abandons us. Every day, we wake up with the need to find affection and use all the strategies and opportunities that come our way to achieve it.

Of course, as in any struggle, we don't always get our way, and we don't always manage to obtain the necessary affection. The differences between individuals in this regard are enormous. While there is a minority of individuals who receive large amounts of affection from the people around them, there is also a large number of people who fail to obtain the necessary affection to thrive adequately.

Affection, as we will see, is a very scarce resource, unlike money, which, thanks to the industrial revolution, we squander without restraint. Therefore, it is very difficult to obtain affection from others, and there are large pockets of emotional poverty in the most economically advanced societies. I can easily imagine that our emotional situation is similar to how we were economically before the industrial revolution when economic misery was the norm. Thus, currently, emotional misery remains the norm. It will be necessary for someone in the future to discover how to launch an "Affection Revolution" so that future societies can have much more affection than we have.

This biological approach to affection allows us to understand, I believe for the first time, the true nature of our emotional relationships and, above all, provides us with much deeper knowledge of the origin and nature of diseases and behavioral disorders, to the extent that it is possible to propose truly effective treatments.

Ways to provide social support

The ways in which humans can support or show affection to people within their groups are very diverse. Human interaction is a multi-channel phenomenon, and many channels are involved in each simple interaction, where both parties are simultaneously emitting and receiving information through them. Despite this complexity, we can distinguish two main ways or channels to provide support, in other words, to work for the benefit of others:

1) Brain work through the exchange of external resources.
2) Brain work through brain-to-brain communication.

Furthermore, for each channel, there are two opposite alternatives:

a) Providing information
b) Taking negative information

Providing information through external resources. On many occasions, people help others by providing some type of material resource such as money, furniture, food, etc. In these cases, it may seem easy to identify a supportive behavior, but we must keep in mind that it is only a supportive behavior if there is some mental work or effort on the part of the provider. For example, if someone needs to get rid of their old television and offers it to their brother, who comes and picks it up, then there is no effort on the part of the person giving the old television, and therefore, there is no support. If the brother benefits, it is thanks to their own effort and they have not received any support. But if a parent buys a new car for their daughter, feeling anxious about potential accidents and making a significant financial effort, then they are providing support (Lin et al., 1985).

The amount of support given is not the economic value of the provided resources, but rather the amount of mental work and effort made by the provider (Uchino, 2004). It's like the story from the New

Testament about the Pharisee compared to the charity of the poor old woman. The value of charity is not its economic value but the effort of the provider. This concept is not easy to understand because we are accustomed to evaluating things by their economic value.

Receiving spent resources. Another way to provide support to others is by taking away their spent materials, the waste. This type of support is mainly given to children and disabled people (Kahn & Antonucci, 1980). Cleaning rooms, clothing, the body, etc., are the most common ways to do it. For example, when parents allow their children to make a mess or break things, they are supporting them in releasing their tension and anxiety.

This type of support is very important and often forgotten. Most parents like to have "good" children, children who behave quietly and respectfully all the time. But this attitude does not support their children; on the contrary, it is to obtain support from them. The "good" behavior of children benefits parents first because they can stay at home peacefully. Conversely, the "bad" behavior of children benefits the children because parents have to put in a lot of mental work and effort to cope with it and cannot stay calm.

Providing brain-to-brain information. The most important way people support each other is by providing information through direct brain-to-brain communication. We use the term "information" in the sense of Shannon (1948), i.e., "reduction of uncertainty." There are many ways to reduce uncertainty through brain-to-brain communication (through verbal and non-verbal communication). Humans are continually exchanging information in their group relationships. We are continually adopting attitudes in response to the behavior of others, and we provide most of the information without saying a word.

In our home, at our workplace, or in public places, we are constantly communicating with other people for nearly 16 hours every day. Our primary channel of interaction and social support is, therefore, direct brain-to-brain communication. Through our attitudes, verbal comments, silences, gestures, or glances, we can boost the self-esteem, confidence, security, or happiness of others. We can motivate others to do what they desire or prevent them from doing something that might be dangerous for them.

But we must realize that, contrary to common belief, this type of support involves a lot of mental work; in other words, it's not free. People often think that making others happy by being kind to them doesn't come at any cost, but that's an illusion because this type of support may not appear to use any material resources. However, when people interact, many physical changes (chemical and electrical) occur inside their brains, even though we can't see them. All these physical changes are the result of the exchange of mental work during each interaction.

For example, think about encouraging a depressed friend. Consider everything you might do to make them feel happy. Perhaps you have to spend many hours to achieve it. In all these situations, you must attend to their needs and forget about your own. You have to think about them and not about yourselves. You need to understand their problems and set yours aside. If you carefully consider this situation, you will realize that instead of feeling happy for helping your friend, you have put in a lot of work and effort for their benefit.

Our daily interactions with others are filled with a multitude of small acts of support throughout the day. Hundreds of these small acts of support occur daily, and most of them involve a small amount of work and effort, but their cumulative effect is substantial. Humans are constantly working with their brains for the benefit of others throughout the day without changing any external resources. Therefore, in the study of support exchange, we must primarily focus on this type of brain-to-brain work exchange.

Absorbing negative information. This form of support is often overlooked and neglected by researchers. Instead of providing positive information, one can support others by receiving their negative information (uncertainty). For example, when a person is anxious and behaves aggressively, you can support them by allowing them to behave in this negative way. Allowing them to release their aggression is also a form of support. When parents allow their children to yell at them, insult them, or hit them, they are supporting them by permitting them to release their tension and anxiety.

It is well-known that one can help others not only by giving advice but also by listening to their problems. In many cases, such as in psychoanalytic therapy, therapists primarily help patients by listening to their problems. It is very common in friendships to listen to friends' problems and attend to their issues. Only by listening attentively can one provide significant support, and this is another way of supporting by taking in negative information from others.

Affective signals

Affection is a need for all social animals, as it refers to the work that an organism performs for the benefit of another. In the evolution of social species towards more complex degrees of social structure, new behaviors have emerged with the function of maintaining the social structure of the species (Trivers, 1971). In the human species, norms, values, rituals, and affective signals have appeared whose function is to maintain the social structure of the group (Durkheim, 1915; Brown, 1991).

Affective signals, in particular, are expressed in a wide range of stereotyped behaviors that are genetically and culturally determined, and their function is to ensure the emotional availability of the sender to the receiver (Eibl-Eibesfeldt, 1986; Harlow & Zimmerman, 1959). Smiles, friendly greetings, signals of acceptance, or promises of support serve to engage the sender and constitute a potential source of affection for the receiver. Both ethology and anthropology extensively study this type of signals or behaviors.

A social organism not only needs the support of its peers in the present but also needs some assurance that this support will continue in the future (Bowlby, 1969). The function of affective signals is to satisfy this need. When one person smiles at another, they are conveying the trust that they can count on them in the future, that they are and will be recognized as a member of their group, and therefore, they are willing to provide affection when needed (Schachter & Singer, 1962). The result is that the person receiving the smile experiences a positive emotion (Ekman & Friesen, 1971).

However, emitting affective signals does not ensure, in all cases, a future provision of affection, as this will depend on the actual capacity of the sender to perform the work required. This explains how, in practice, individuals who emit affective signals (smiles, greetings, promises, etc.) may later be unable to provide the required assistance because they do not have the necessary capacity to do the work. This mismatch between emotional intention and actual emotional capacity leads to frequent and varied conflicts in human relationships (Deutsch, 1949; Kelley & Thibaut, 1978).

Finally, affective signals are also a way to encourage reciprocity in emotional exchange, as the receiver of these signals feels an obligation to reciprocate the potential affection received (Batson, 1991). If an organism that performs work for the benefit of another, in other words, provides real affection to the other, does not emit affective signals, there is a risk of not being compensated by the other (Trivers, 1971). Thus, we not only help others but also let them know we are helping them so that the social mechanisms (genetic and cultural) responsible for establishing reciprocal exchange can operate (Gouldner, 1960).

The affective balance

Since the provision of social support benefits the recipient and harms the provider, and because humans engage in a constant exchange of social support, it is necessary to introduce the concept of the balance of social support exchange or affective balance.

The affective balance is the difference between the social support received and the social support given by an individual (person, group, etc.) during a certain period of time. Thus, a positive affective balance means that the individual has received more social support than they have given during the specified period of time. Given the complexity of continuous exchanges of social support, the affective balance conceptually provides us with an overall summary of these exchanges for a given individual and period of time.

As we have seen, groups are based on the exchange of support among their members, and supporting others involves working for their benefit. In every group relationship, there is an exchange of effort (free energy) for the benefit of the other, in other words, an exchange of support. Therefore, it is of interest to determine the balance of this exchange, whether it is balanced or unbalanced. For example, in a husband-wife relationship, how much work does the husband do for the benefit of his wife, and how much work does the wife do for the benefit of her husband? The difference represents the work equilibrium in that specific relationship.

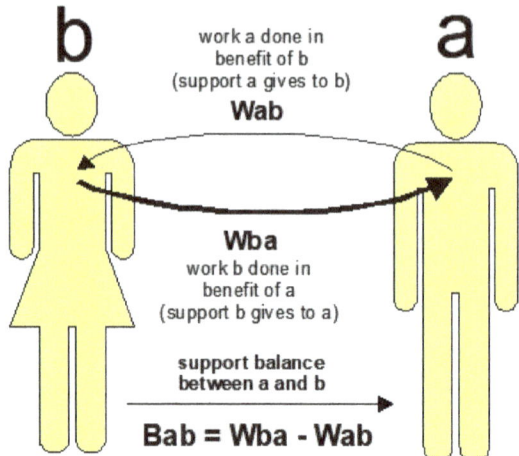

Figure 9. This figure illustrates the work performed by two individuals in a social support exchange situation and the balance of social support between these two individuals.

If W_{ab} is the work done by a for the benefit of b, and W_{ba} is the work done by b for the benefit of a, the work balance between a and b (with respect to a) is

$$B_{ab} = W_{ba} - W_{ab} \qquad (35)$$

If the balance B_{ab} is zero, both sides exchange the same amount of support. If the balance is positive, then a receives more support from b than a is giving to b, and vice versa.

Of course, the balance with respect to b is opposite to the balance with respect to a.

$$B_{ba} = W_{ab} - W_{ba} \qquad (36)$$

If a relationship is balanced, both benefit because neither of them loses energy in this particular relationship. The effort one made for the other is compensated by the work the other did for one. But what could happen if the balance is very imbalanced? What can happen to the member whose balance is negative?

For example, let's suppose that in a husband-wife relationship there is

$$B_{hw} <<< 0$$

In other words, the work balance between the husband and wife, with respect to the husband, is highly negative. This means that, on average in the relationship, the husband's brain works much harder to support his wife than his wife's brain works to support him. The consequences of this unequal exchange between them can be significant or not, depending on the situation of their other group relationships.

With the cerebral work balance defined for a specific relationship, we could define the total cerebral work balance that a subject has across all their relationships. If a subject a has established n relationships during a certain period of time, their total cerebral work balance is equal to the sum of all the balances across their n relationships, i.e.,

$$B_a = \sum_{n=1}^{i} B_{ai} \qquad (37)$$

B_a expresses the overall gain or loss of free energy for subject a across all their relationships, and this measure has important consequences for the survival and well-being of the subject.

Relationship between affective balance and health [63]

The significant development of medicine from the late 19th century to the present day has completely transformed the quantity and quality of human well-being, especially in advanced industrial societies (Porter, 1997). Pasteur's discovery (1861) about microscopic life and its tremendous impact on the health of living organisms has led to significant control of infectious diseases (Dubos, 1959). However, alongside the control of these diseases, we are witnessing the emergence of a large number of illnesses that had little chance of appearing before the 20th century (Omran, 1971).

As we enter the third millennium, we find ourselves in a paradoxical situation regarding our health. Despite a significant increase in healthcare spending and medical research since the last century (Anderson & Poullier, 1999), we are not experiencing a significant reduction in the incidence of diseases in our lives and must accept a growing proliferation of many of them (Barton & Husk, 2000).

The common characteristic of all these 'new' diseases is that they are not caused by microbial agents, meaning they are not caused by viruses or bacteria (Frenk et al., 1991). Diseases such as cancer, heart attacks, allergies, depression, or obesity leave the scientific community perpetually puzzled about their origins (World Health Organization, 2005). We know many things about them, how to alleviate their symptoms, and even how to treat them, but their causes remain a scientific mystery today (World Health Organization, 2002).

---

[63] The texts in this unit were written in the year 2002 and are part of the document "Social support, health and well-being." (Editor's Note)

In this section, I want to explain the most important discovery I have made in my years of research:

Chronic affective deficit causes non-infectious diseases and behavioral disorders.

By understanding that affection is a biological resource, primarily neurological, I realized that systematic affective deficits directly cause neurological dysfunctions and indirectly somatic diseases.

You must understand that a neurological dysfunction is any alteration in brain behavior that implies harm to the organism's survival. Neurological dysfunctions can manifest as psychological disorders (depression, obsession, distress, dizziness, phobias, etc.), risk behaviors (drug use, stress, inadequate nutrition, etc.), developmental deficits (academic or occupational failure, etc.), or violent behaviors (abuse, homicides, robberies, etc.). Both psychological disorders and risk behaviors eventually lead to somatic diseases that are not caused by viruses or bacteria (cancer, heart attacks, embolisms, ulcers, pains, allergies, injuries, suicide, etc.).

However, the specific symptoms caused by an affective deficit depend on various factors, with genetic (genetic predisposition), cultural (norms, values, prejudices, knowledge, etc.), and environmental factors (economy, social status, geography, employment, social relationships, etc.) playing a significant role in determining the symptomatic configuration in each particular case.

In summary, I believe that the cause of the majority of non-communicable diseases and brain disorders that humans suffer from is due to the systematic lack of affection. All of this allows for the definition of a new scientific framework for Psychology as a biological discipline, following the path opened by Medicine in the 19th and 20th centuries.

Non-communicable diseases

Despite the great technological advancements, modern humans continue to suffer from illnesses as much or more than their ancestors. Pasteur's discovery allowed us to understand diseases caused by microorganisms (bacteria and viruses) that attack the human body. Although it can be said today that such diseases are under control, a large number of diseases have emerged of which their causes are unknown (cancer, hypertension, depression, bulimia, etc.).

Currently, medicine divides acquired diseases into two major groups: communicable and non-communicable diseases.

Communicable diseases are infectious diseases caused by the attack of another living being. In the struggle for survival, many species can obtain the resources they need from our bodies and achieve their reproductive success (Ewald, 1994). Although evolution has allowed us a certain capacity for defense, in most cases, we are unable to win the battle (Nesse & Williams, 1994). This type of diseases is well understood. We know their causes, and in most cases, we know how to prevent and cure them. Some examples of human infectious diseases are polio, pneumonia, influenza, diarrhea, enteritis, tuberculosis, diphtheria, meningitis, cholera, or AIDS.

The other main set of diseases is non-communicable diseases. These are not caused by the attack of bacteria, viruses, fungi, protozoa, or parasitic worms. The cause of this type of diseases is still unclear in most cases, especially in late-onset diseases. Some examples of these diseases are cancer, heart attacks, cerebrovascular diseases, depression, diabetes, nephritis, or anorexia.

In recent decades, the idea that psychological and social factors should play some role in most non-communicable diseases has developed in the medical community (Engel, 1977; House, Landis & Umberson, 1988; McEwen & Stellar, 1993). This idea arises from two main issues:

1) Despite extensive research, the causes of major non-communicable diseases remain unknown. This situation has forced scientists to seek answers in areas other than traditional ones.

2) The daily experience of doctors leads them to suspect that most non-communicable diseases are related to the patient's lifestyle (Marmot & Wilkinson, 2001). It is common for patients to visit the doctor and, after a comprehensive review, the doctor tells them that they are fine, and the cause of their condition comes from "nervousness," stress, anxiety, etc.

Some doctors have tried to reflect this way of thinking in what is called the "biopsychosocial model" (Engel, 1977). But none of the trials conducted so far has been able to establish a clear and direct relationship between psychosocial factors and non-communicable diseases. To date, the main approach has been to find some kind of epidemiological correlation between non-communicable diseases and behaviors, habits, or environments (Berkman & Kawachi, 2000). All of these are statistical studies that show statistical correlations. However, no coherent and explicit theory has been developed.

We know that smoking, drinking alcohol, or excessive consumption of certain foods correlate with some non-communicable diseases (Doll & Peto, 1981; Stampfer et al., 2000). However, all studies conclude that the exact percentage of people who develop these diseases depends on many factors, including the amount and duration of exposure to toxic materials and individual characteristics such as age, genetics, and lifestyle (Rothman, Greenland, & Lash, 2008; Willett, 2002). Therefore, these percentages can vary from one study to another, meaning that although we have many statistical relationships with risk factors or potentially toxic materials, we know almost nothing more.

This does not mean that we know nothing about these diseases. Modern medicine can alleviate many of their symptoms and provide a good life expectancy for most patients. Furthermore, we know many aspects of the biochemical processes involved in them, but the problem I want to highlight here is that the causes of many non-communicable diseases remain unknown at this time.

This situation contrasts greatly with what we know about infectious diseases. Since the mid-19th century, a coherent biological theory has been developed for this type of disease (Dubos, 1965). Today, we know that some microorganisms (bacteria, viruses, and fungi) and some animal parasites are responsible for diseases such as tuberculosis, pneumonia, enteritis, nephritis, bronchitis, diphtheria, etc. In the 20th century, communicable diseases had undergone a drastic reduction in developed countries because we know their cause and are more capable of finding efficient methods to prevent and cure them (Cutler & Miller, 2005).

In short, the focus of medical research to find the chemical mechanism involved in any non-communicable disease makes us forget how closely related any organic event is to the whole organism and its environment. Focusing on the inside of the cell, the inside of its nucleus, or the inside of its genes makes us forget that an organism is a sophisticated society where all its components are closely related in a cooperative relationship.

For example, the relationship between smoking and some cancers forgets that there is a complete organism among them. Tobacco does not directly access the cells, meaning that there is not only a direct interaction between cells and tobacco. For tobacco to enter the cells, the organism needs to find cigarettes, light them, and smoke them. In other words, a complex process of individual events is required before tobacco can access the cells. Prior and complex behavior is needed for tobacco to enter the cells. And although doctors talk about behavior as a risk factor for non-communicable diseases, there is practically no research at that level of events.

The disease and its causes

In the first part of the book, you have been able to see that living beings are open thermodynamic systems that need to interact with the environment to survive (Schrödinger, 1944). Living beings have to reduce their entropy or uncertainty to the environment to stay away from thermodynamic equilibrium, that is, to stay informed. However, since the interaction of living beings with the environment consists of increasing disorder by 'feeding' on the order of the environment, the ease or difficulty of carrying out this interaction will depend on the characteristics of the part of the environment with which the system interacts.

As we have seen, the work of living beings to obtain resources from the environment focuses on other living beings in their environment because they are the only potential recipients of entropy they must eliminate, or in other words, they are the only ones who possess the necessary resources to survive (Odum, 1983). However, since they are also living beings, they resist with all their might absorbing the entropy of others, or in other words, giving up the resources they possess, thus establishing a struggle for entropy or resources, in other words, a struggle for survival.

Ecologists analyze interactions between living beings based on the exchange of resources that occurs between them, considering a resource to be any substance, space, position, status, role, etc., that contributes to the survival and reproduction of an individual (Begon, Townsend, & Harper, 2006). In my analysis of the behavior of living beings, the concept of a biological resource is identical to the physical concept of negentropy and information, in other words,

$$Resource = Negentropy = Information$$

This struggle for resources (negentropy, information) gives rise to different types of relationships or interaction strategies among organisms fighting for survival. In this constant struggle, living beings can only stay healthy if they manage to transfer their entropy (uncertainty) to the environment, in other words, if they obtain enough resources from the environment (Schneider & Sagan, 2005; Schrodinger, 1944; Kaufmann 1993). Only when there are difficulties in this transfer will a living being eventually become ill. Therefore, any kind of illness or pathology is ultimately a consequence of the resistance put up by the environment of the sick individual to receive the entropy that the individual needs to transfer. Illness (whatever its nature) and ultimately death are states in which the entropy of the living system significantly increases, in other words, entropy (disorder) is the origin of illness and death.

In infectious diseases, viruses and bacteria from the environment transfer entropy to the patient by feeding on their resources. However, as we have discussed, there are many diseases and pathologies that are not caused by viruses or bacteria (non-communicable diseases) such as heart attacks, cancers, obesity, anorexia, hypertension, allergies, Alzheimer's, osteoarthritis, etc.

In the medical community and popular thought, the idea has taken hold that such diseases occur "spontaneously" or by spontaneous generation. At most, their causes are attributed to the behavior of the patient (smoking, drinking, lack of exercise, diet, stress, etc.), and in no case is it considered that the causes may be outside the patient. On the other hand, there is a growing awareness among health professionals that affective phenomena significantly influence the course of such diseases, although they may not specify the type of influence (House, Landis & Umberson, 1988; Antonovsky, 1987; Segerstrom & Miller, 2004).

I believe that if we can understand the fact that affection is a biological resource, with all the consequences that follow from it, we will be in a position to take a big step towards understanding the causes of all these diseases (Cohen & Wills, 1985; Uchino, 2006). There is now enough evidence to believe that affection directly affects the brain's capacity to work, and this is fundamental to understanding its enormous impact on all kinds of diseases and disorders (Davidson & McEwen, 2012; Sapolsky, 2004). The brain is not just another organ of the body; it is the central unit that governs the fate of each and every part and function of the entire organism. Therefore, its health is directly dependent on the health of the brain itself (Gazzaniga, 2005). A brain that cannot perform its functions correctly will eventually induce organic dysfunctions in any other part of the organism.

Brain work is the most important activity carried out by humans. First, the brain has to control our metabolic functions to maintain the dynamic balance of the body, primarily by acting on our glands. Although we are not aware of this work, neurology and medicine have proven its existence (Kandel, Schwartz, & Jessell, 2000; McEwen & Lasley, 2003). Second, the brain has to control our muscular work to provide the resources we need to survive and support others. Our muscles, without proper brain guidance, cannot obtain the resources we need (Wolpert, Doya, & Kawato, 2003). The brain has to control many environmental variables as well as internal variables to guide our muscular work correctly. Third, the brain has to obtain information in order to function properly. The brain is like a computer. It has hardware devices and software programs. Its performance depends on the quality of both elements. But since hardware devices are provided by our DNA and maintained by our metabolism, software programs must be obtained from other brains. The more information (data and software programs) a brain has installed, the more work it can do (Pinker, 1997).

Aware of this situation, one can realize that anything that interferes with brain function can cause metabolic, behavioral, and informational dysfunctions in humans (Kolb & Whishaw, 2009; Lupien et al., 2009). This means that our health and well-being depend directly on our brain's performance. Metabolic dysfunctions like cancer or heart attacks, behavioral dysfunctions like drug use or traffic accidents, and information dysfunctions like academic failure or depression are directly related to poor brain performance.

Therefore, it is of vital importance to find the causes that interfere with and reduce brain performance in order to understand and prevent conditions such as cancer, traffic accidents, depression, obesity, etc. As I will detail in the upcoming chapters, I believe we can find these causes in the imbalanced exchange of social support in social relationships.

The affection deficit

What is an affection deficit? We have seen that affection is the social support that humans exchange in order to survive, and this is done through unpaid work for the benefit of others. Each person receives help (affection) and, in turn, provides help (affection) to others. At the same time, each individual has different affective needs, in terms of quantity and quality, depending on their level of autonomy. For example, children need large amounts of affection because they have very little capacity to obtain the resources they need on their own. Adults, on the other hand, generally need less affection, although they cannot do without it (Bowlby, 1988; Reis & Collins, 2000; Taylor & Stanton, 2007).

When a person lacks sufficient help to survive adequately, they experience an affection deficit. But this not only takes into account the help received but also the help provided. If a person provides much more help than they receive from others, they can also experience an affection deficit (Ingersoll-Dayton, Morgan, & Antonucci, 1997). Therefore, when the difference between the received and provided help is less than a person's affective needs, an affection deficit occurs (Cohen et al., 2000; Liang et al., 2001; Thomas, 2010).

In the case of children, an affection deficit generally occurs because they do not receive sufficient help to develop normally (Bowlby, 1988; Rutter, 1981). However, it is important to note that this lack of help can occur both due to underprotection and overprotection. Underprotection means that the child has to face problems without having the necessary capacity to overcome them, leading to an imbalanced development of their abilities and personality (Egeland & Farber, 1984). On the other hand, overprotection means that the child does not acquire the necessary skills to survive, resulting in a serious developmental deficit, making them incapable of facing life's challenges later on (Clarke-Stewart, 1988). Helping a child's development means protecting them from situations they cannot overcome and exposing them to situations they can handle (Sroufe, 1997). The "advantage" children have is that their youth largely hides the health impairments since they possess great vitality, although they eventually appear sooner or later (Halfon & Newacheck, 1999).

In children, we often intuit the affection deficit, but in adults, it usually goes unnoticed. In their case, the affection deficit is usually caused by providing help to others beyond their capabilities (Liang et al., 2001; Pines, 1996; Maslach & Leiter, 2008). Adults have a greater capacity for affection, and therefore, the deficit occurs when the help they provide to others deprives them of the energy needed to survive. People who tend to help others without expecting or receiving any kind of reward often experience an affection deficit.

An affection deficit causes excessive stress on the brain because it either has to deal with too many situations it is not yet prepared to handle in the case of children, or it has to deal with too many other people's problems while neglecting their own problems in the case of adults. In general, every adult can provide a certain amount of help without their brain being unable to meet the demands of their own survival. However, there are many circumstances that can lead an adult to unknowingly exceed their personal limit of helping others and suffer from an affection deficit.

The affection deficit as an underlying cause

Nowadays, it is accepted that the majority of non-communicable diseases are closely related to what is referred to as "environmental factors," but what about factors related to human relationships? In other words, can most non-communicable diseases be related to social relationships? Can some kind of parent-child relationship cause cancer in a child? Can some kind of husband-wife relationship trigger a heart attack in the husband? Can some kind of mother-daughter relationship lead to depression in the daughter?

As we have seen, all helping behaviors involve the complete or partial resolution of someone else's problem, and this always involves the use of brain work, as well as muscular or economic effort in many cases.

The brain, along with the entire nervous system, is the organ that animal species possess to direct their behavior in the direction of survival and, in the case of the human species, achieving higher levels of well-being. It is a computational organ that continuously evaluates the situation (perception) and makes decisions (motor and glandular action) in accordance with the goals it pursues. However, its computing capacity is limited based on the information it stores, and this varies in each individual.

Just as when a computer is given too many tasks, it starts to slow down and can even "crash," the brain experiences something very similar. To simplify, we can think of the brain as a machine for solving problems (related to survival and well-being). When the brain faces a quantity of problems for which it is prepared, it makes the most accurate decisions. But if the demand increases to the point of surpassing its computing capacity, it will start to make some less accurate decisions.

This is the underlying cause of the relationship between affection deficit (negative affective balance) and deteriorating health. When a person becomes entangled in the problems of others (usually those close to them), their brain's computing capacity can easily be overwhelmed. If this situation persists, their brain will systematically make erroneous decisions that will ultimately negatively affect their survival and well-being, that is, their health.

This situation of becoming entangled in the problems of others is more common than one might think. Social structure is largely based on the help that individuals can provide to others, and therefore, it is something positive. However, not all individuals have the same capacity to help others, that is, to process the problems of others in addition to their own. In this regard, there are significant individual differences.

Social structure is built on cooperation, that is, the exchange of affection among its members. However, individuals with lower affective capacities try to compensate for this deficit by seeking help from people close to them, whether or not those individuals have the capacity to provide it. On the other hand, individuals with greater capacities tend to provide help, driven by the social values of cooperation. Finally, individuals with limited helping capacities are often forced to provide help due to strong emotional bonds with certain people in their environment (parents, grandparents, partners, children, friends, colleagues, etc.). All of this frequently leads to situations in which a person finds themselves offering help to others beyond their capabilities, and consequently, their brain, unable to process the volume of problems it is handling, consistently makes erroneous decisions that, in most cases, affect the individual themselves.

Although it is unknown to what extent our brain can control every event in our body, it is well-established by neuroscience and medicine that the brain exercises broad control not only over our behavior but also over the functions and metabolism of our body. The brain is a central computer that controls and directs our entire life, from chemical events within our body to our most complex behavior (Kandel, Schwartz, & Jessell, 2000; Squire, et al., 2003).

Therefore, if an affection deficit persists and the brain does not have sufficient capacity to correctly evaluate each situation, it will start to process vital information incorrectly (Sapolsky, 2004; McEwen, 2007). This leads to an increase in brain inefficiency (neuronal dysfunction) and the consequent emotional errors: it believes it is hungry when it is not, it believes there is no danger when there actually is, it doesn't have time to think about itself, or it doesn't care about the harm caused by smoking, and so on. The result of this persistent inefficiency is the appearance, sooner or later, of some form of brain disease or disorder.

This is why in recent years numerous studies have been demonstrating the correlation between deteriorating health and the affection deficit. Some studies have shown a correlation between the affection deficit and certain metabolic dysfunctions or non-communicable diseases. Various authors have linked the onset of type 2 diabetes with the affection deficit (Hawkley & Cacioppo, 2010; Knol et al., 2006; Nonogaki et al., 2007; Pouwer, Kupper, & Adriaanse, 2010; Umberson & Montez, 2010). Other research has demonstrated that the affection deficit is related to the occurrence of cardiovascular diseases and coronary artery diseases (Piferi & Lawler 2006; Rosengren et al., 1991; Rosengren et al., 2004; Barth et al., 2010; Orth-Gomér et al., 2000). It has also been shown that a lack of social support is significantly associated with an increased risk of impaired kidney function (Boulware et al., 2005; Jansson et al., 2012; Palmer et al., 2013). Some studies have also shown the correlation between the affection deficit and the incidence and survival of cancer (Chida et al., 2008; Fagundes et al., 2014; Lutgendorf et al., 2011; Pinquart & Duberstein, 2010). Furthermore, there are numerous studies that have demonstrated the relationship between the affection deficit and various behavioral dysfunctions such as drug consumption (Goings et al., 2019; Hawkins et al., 1992; Sinha, 2008; Valkenburg & Peter, 2007), academic failure (DeRosier et al., 1994; Suldo et al., 2008; Rueger et al., 2010), reckless driving (Lagarde et al., 2004; Useche et al., 2018), as well as its relationship with numerous neurological disorders such as depression (Cacioppo et al., 2006; Santini et al., 2015), distress (Kawachi & Berkman, 2001; Segrin & Passalacqua, 2010), anxiety (McHugh et al., 2011), anorexia (Troisi et al., 2006), attention deficit (Mikami & Hinshaw, 2006), etc.

In summary, the health of the body depends directly on the accuracy of the decisions made by the brain, and this accuracy depends on both the brain's capacities and the workload it bears (Liang et al., 1999; Lupien, et al., 2009; McEwen, 1998; Sapolsky, 2004). The help of others can compensate for the lack of the brain's own capacities, but a continued excess workload inevitably leads to erroneous decisions and a deterioration of health (Berkman, et al., 2000; Cohen & Wills, 1985; House, Landis & Umberson, 1988; Uchino, 2006; Holt-Lunstad et al., 2010).

Now, we need to ask, what situations can cause the brain to be unable to perform its function adequately? There are two clearly distinct situations.

First, when the brain itself lacks the necessary capacities to perform its task. To carry out its function, the brain needs a lot of information (skills, knowledge, abilities, values, attitudes, etc.). Some of this information is received through genetic inheritance, and another part is received from peers in the form of affection through learning. Thus, either due to genetic mutation or a low level of received affection, the brain can find itself unable to successfully carry out its function.

Secondly, dysfunction can also occur when the brain, despite being fully capable of doing its job, has to help other brains beyond its capacities. In this case, the brain fails due to an overload of work caused by the excessive transfer of affection to other brains. In other words, the main situation that can cause a cerebral workload overload, under normal conditions, is the excessive provision of help to people in close proximity, which is very common in our social environment.

From this, we can derive our hypothesis, namely, that

A negative affective balance, that is, providing assistance to others beyond one's own capacities or a lack of social support from others, leads to health deterioration.

With this approach to the affective function, we have established a theoretical framework that allows us to understand the relationship between affective exchange and human health, thereby achieving the ultimate goal of all psychology: promoting the health and well-being of human beings.

We don't just get sick due to the loss of resources in favor of certain microorganisms (Pasteur, 1861). We can also get sick due to an affection deficit, whether it is caused by the excessive transfer of affection (resources, free energy, work) to our peers or by insufficient help received. Just as in communicable diseases, an external agent causes non-communicable diseases. While in communicable diseases, microorganisms feed on our resources and lead to an increase in relative entropy, in non-communicable diseases, our peers feed on our resources and generate an increase in relative entropy. By taking advantage of the affection we provide them, our peers obtain our free energy, use our cerebral capacity, and thus degrade us and generate entropy.

This relationship between affective exchange and health materializes in two fundamental facts: 1) The extreme dependence of organic functions on the proper functioning of the brain, and 2) The weight of affective work falls much more on the brain than on the muscles in industrial societies. Although it may seem surprising, this relationship between excessive transfer of affection and health deterioration is beginning to be demonstrated in various research studies (Liang et al., 2001; Piferi & Lawler 2006; Pinquart & Sörensen, 2003; Schulz & Sherwood, 2008; Seltzer et al., 2001).

Mental work and survival

Now we have the elements to establish a direct relationship between social relationships, illnesses, and well-being through the concept of affection as cerebral work. These relationships are represented in Figure 11.

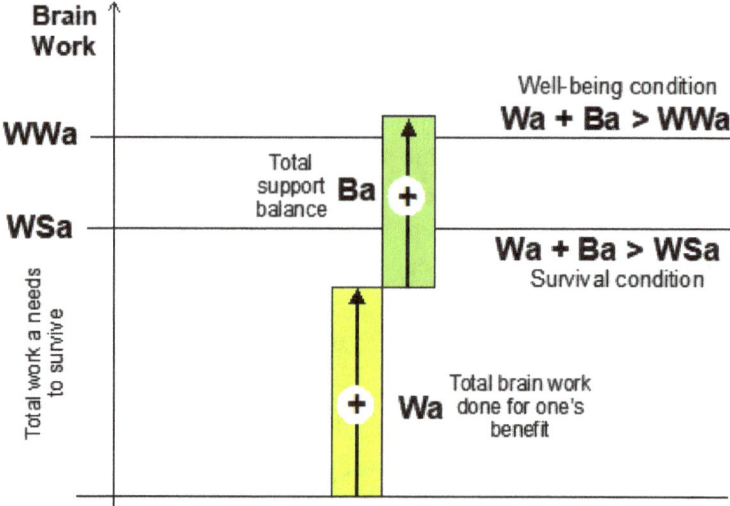

Figure 10. Relationship between group relationships, disease, and well-being through cerebral work.

Along the vertical axis, the quantities of cerebral work are represented. Then, we define two threshold lines, WS_a and WW_a, which represent the amount of cerebral work required for survival and the amount of cerebral work required for well-being, respectively.

In other words, WS_a represents the minimum cerebral work required to avoid death, and WW_a represents the minimum cerebral work required to avoid being ill. Between these two lines lies the region of illness, where one could survive but be unwell.

In Figure 11, we can see a case in which an individual is unable to perform all the cerebral work required for both survival and well-being, such as children and the elderly, i.e.,

$$W_a < WS_a$$

But this individual could survive and enjoy well-being due to a positive balance of support (mental work) in their relationships. From their relationships, they receive enough support that allows them to live in well-being, even if they are not capable of achieving it on their own.

In other words, in this case, the health and well-being of this individual are caused by a positive balance of cerebral work from all their relationships. For example, if it were a 6-year-old child, we are representing that they are growing up in health and well-being thanks to the mental work they receive from their parents, educators, etc. Let's explore other key cases in Figure 12.

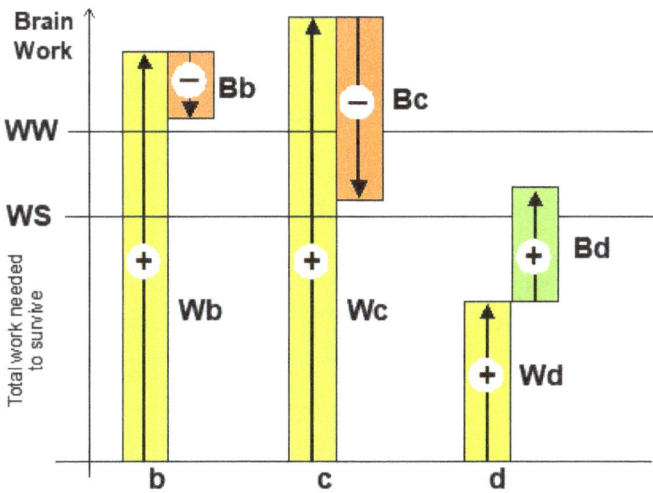

Figure 11. Key cases where we observe how the brain's lack of resources to work on its own well-being leads to illnesses.

Subject **b** has a high capacity for cerebral work.

$$W_b > WW$$

Therefore, they are capable of performing all the work necessary to live in well-being. Although they have a negative cerebral work balance in their relationships ($B_b < 0$), it is not negative enough to endanger them.

$$W_b + B_b > WW$$

Subject **c** is in the region of illness.

$$WS < W_c + B_c < WW$$

Not because they don't have enough cerebral work capacity ($W_c > WW$) but because they have a very negative cerebral work balance among all their relationships.

$$B_c << 0$$

This individual is providing much more support to their peers than they are receiving from them, and although they could live well thanks to their own mental work, they are ill and could die if the situation does not change. Finally, subject **d** is also in the region of illness.

$$WS < W_d + B_d < WW$$

but for a different reason. They do not have enough cerebral work capacity to survive on their own.

$$W_d < WS$$

and they do not receive enough support from their peers to live in well-being, so they remain ill.

$$B_d > 0$$

There are many other possible situations, but I just want to demonstrate how to handle these concepts to achieve an understanding of the close relationship between social relationships and health and well-being through cerebral work.

The region between *WS* and *WW* represents the dysfunction of the brain to control and direct all events involved in the well-being of the individual correctly. Some dysfunctions can affect metabolic processes, while others can impact behavioral success and information achievement. But, ultimately, we can formulate our theory as follows:

The ultimate cause of cerebral dysfunction in maintaining the well-being of the individual is that there is not enough available cerebral work capacity for the individual due to two possible sources: 1) their own capacity for cerebral work and 2) the balance of cerebral work among all their social relationships.

In summary, relationships within human groups require greater attention to prevent most modern diseases and behavioral disorders. We live in a culture that doesn't prioritize our relationships because we believe it's a natural disposition in human beings and that they have no influence on our health and well-being. In this research, I have tried to argue, perhaps in an abstract manner, how important human support exchanges are in each of our relationships. Ultimately, I have attempted to demonstrate how crucial it is to consider the economy of our support exchanges when understanding the origins of non-communicable diseases.

Affective predation

Predation is the fundamental relationship among living beings because it constitutes the most primitive way of obtaining resources from other living beings (Krebs & Davies, 1993; Begon, Townsend, & Harper, 2006; Pianka, 2000). Predation proper consists of a very intense interaction in a very short time between an individual A (predator) that inflicts great harm or death upon B (prey). The predator obtains so many resources from the prey that it can no longer sustain its life. However, in addition to predation proper, there are other types of predation relationships in nature.

Parasitism is a predatory interaction in which an organism (parasite) continuously obtains resources over time from another organism (host), gradually weakening it until it causes disease and death. There are many variations of parasitism, but in general, the parasite lives off the effort made by the host. The host must obtain resources for itself and for the parasite, making the burden much greater. The parasite is not interested in the death of its host, but sooner or later, the host becomes ill and eventually dies, forcing the parasite to find a new host to survive. This does not mean that the parasite does not have to fight; it has to fight to prevent the host from getting rid of it, to avoid detection, and to maintain the relationship. Parasitic relationships are extremely abundant in nature, and the ways in which the parasite achieves its purpose are wonders of biological engineering (Combes, 2001; Poulin, & Morand, 2000).

If we properly analyze the relationships of anyone suffering from a non-communicable disease, we will easily find affective relationships in which they are the host of one or several affective parasites. We will find a parent, a mother, a spouse, a friend, or a coworker who systematically obtains affective resources from the ill person.

Affective parasitism in human relationships is extraordinarily abundant and occurs with economic, affective, and knowledge resources (Hawley, 1950). However, affective parasitism is the most common. Affective parasitism occurs in most intimate or familial relationships. The relationship between the parasite and the host is stable because the parasite obtains moderate amounts of affection from the host, allowing the host to continue to survive and recover. However, if the relationship persists, the host eventually becomes ill and dies because it is not possible to bear this burden for so long.

The vast majority of affective deficits are caused by parasitic relationships of the ill person (host). In this sense, we can affirm that most acquired diseases are caused by predation relationships to which the sick person is subjected, whether they are relationships with microorganisms (infectious diseases) or with organisms of the same species (affective relationships). As I mentioned earlier, while in infectious diseases, microorganisms are responsible for the deterioration of our health because they feed on our resources, in non-communicable diseases, it is the people in our environment who are responsible for the deterioration of our health because they take advantage of our resources.[64] In both cases, we are dealing with a predatory relationship in which one organism obtains resources from another organism that is thereby harmed.

Of course, a healthy organism doesn't become ill due to occasional lapses in affection. When a person is subjected to a lack of affection for a long time (years), or receives degraded affection from others for an extended period, they become ill in one way or another, depending on various concurrent variables, including, perhaps most importantly, genetic predisposition to a particular disease. However, what is important to highlight here is that it is the parasitic relationships with

---

[64] Let's remember that, in the context of the present theory, the concept of biological resource is equivalent to the concepts of negentropy, negative entropy, information, or free energy. (Editor's Note)

our loved ones, the relationships that persist over a long time, that make us ill due to continuous affection deficit.[65]

In general, the sick person is often a generous, noble, and kind individual with the people around them on a daily basis. They are usually someone who always tries to encourage those in need, finds it difficult to say 'no,' and anticipates the desires of their loved ones. The concept of affection and affection balance allows us to understand that affection deficit occurs not only from not receiving affection but also from providing it excessively. This is one of the most important discoveries from my point of view. Not only do those who receive little affection become ill, but also those who provide it in excess. Extremely kind, responsible, or helpful individuals often become ill, such as those who care for others (Schulz & Sherwood, 2008). This also clarifies a well-known fact: professionals dedicated to caring for others, such as doctors, nurses, teachers, police officers, or psychologists, often suffer from health problems more frequently (Figley, 1995; Pines & Aronson, 1988).

The person who tends to give affection to others or, more precisely, the person who gives more affection to others than they receive in return, will eventually become ill in a way determined by multiple genetic, cultural, and social variables. Thus, we can observe that people who die at a younger age are those who presented themselves wonderfully to others, meaning they gave large amounts of affection. Children who die are almost religiously remembered because they were like 'angels' to those around them.

It could be argued that people who have dedicated their lives to helping others have enjoyed very good health and great longevity. I can only say, following my hypothesis, that they either had extraordinary abilities (very unlikely) or, despite helping others a lot, they also received a lot of help from people in their environment,

---

[65] It is true that there is a clear genetic influence in determining which disease is more likely to occur, but the cause of the disease, its trigger, is not the gene but the patient's affective relationships. In most non-communicable diseases, the gene determines the location and form of the disease, along with the patient's culture, but the disease is caused by some form of affective predation on the patient.

allowing them to maintain a balanced or even positive affection balance.

In fact, people who, in their affectionate relationships, do not attend to the needs of others, always demand and obtain the attention of others or impose their needs and desires over those of others, never become seriously ill and survive to very old ages. People who die at a very advanced age are those whose behavior towards others was not positive. Demanding and tough individuals, authoritative and selfish. Although there are cases of people who, even though they died at a very advanced age, are not remembered with bitterness, this is due to their excellent camouflage. The principle of the struggle for life, and for affection in particular, requires that the maintenance of life necessarily involves causing death around it. Thus, young people who die are remembered as sources of life, while people who die very late have caused death and much pain in their closest social environment.

In reality, the casuistic complexity is infinite, and it is not possible to generalize relationship patterns since each case is different from the others. What they all share, however, is the existence of predatory relationships against the sick individual. In the last part of the book, we will see that in industrial societies, every romantic relationship evolves into an affective parasitic relationship of one on the other. We will also see how grandparents become affective parasites of their children, generally causing them serious harm. Finally, we will see how children must be parasites (of affection and other resources) of their parents if they want to survive since they do not have the capacity to obtain resources on their own. In conclusion, we will see how the majority of affective relationships in humans are parasitic relationships because one often takes advantage of the initiative and capacity of the other.

Perhaps the problem we have in recognizing this situation is the significant amount of time required for a person to become ill due to a lack of affection. When someone becomes ill, we tend to look for its causes in specific and immediate events, and we find it challenging to appreciate long-term deficits in affection. I believe that in future generations, this will become more evident due to their increasing genetic vulnerability. Since there is no genetic selection in modern societies, organisms become dramatically weaker and less resistant to precarious conditions. Thus, it is expected that in the future, the incidence of affection deficiency will become more apparent, and its effects will manifest more quickly.

I understand that this theory is very difficult to comprehend and even more challenging to accept. Unfortunately, when we examine the facts objectively, as they are, I believe that in the end, no one will be able to deny that things are as described. This is made more challenging for us because we are strongly influenced by utopian ideas about our existence. Ideas that are undoubtedly beautiful and that we all wish were true, but for those who scientifically study human beings, they are nothing more than fantasies and good intentions that lack veracity.

However, this harshness in the real view of facts is only one side of the coin. Fortunately, there is another, much more interesting side. To the extent that we deeply understand our affective relationships, we will be able to manage affective resources much more effectively and, therefore, achieve a higher overall quality of life.

The symptomatology of affective deficit

Although an affective deficit can cause many diseases, it doesn't determine the specific form these diseases take. The particular symptoms caused by an affective deficit depend on several factors, including genetic factors (genetic predisposition), cultural factors (norms, values, prejudices, knowledge, etc.), and environmental factors (economics, social status, geography, employment, social relationships, etc.) that determine the symptomatic configuration in each particular case.

This complexity is due to the brain's enormous complexity and its central role in the functioning of the entire organism. A brain dysfunction can affect any bodily function in a multitude of ways. The combinations are nearly infinite, and as a result, the symptoms can be highly diverse. Since it's impossible to unravel the structure of the information stored in the brain, we can only approximate it through external elements that configure it.

Simplifying, we can say that the brain receives three basic types of information: first, genetic information that is determined by the specific nature of the organism it resides in, including itself (information about the 'hardware'). The brain has to control a vast number of genetically defined organic variables (heart, metabolism, stomach, blood circulation, bones, muscles, etc.). Secondly, the brain must operate with cultural information, which reaches its highest expression in the human species. Knowledge, values, social norms, or symbols constitute very complex information that operates directly within and from the brain (action programs or 'software'). Finally, the brain must process a significant flow of environmental information determined by the external conditions in which the organism must act. The interrelation and integration of these three modes of information in each particular brain determine the specific way in which brain dysfunctions manifest in that organism. Thus, we can talk about the simultaneous and variable incidence of these three factors in determining the particular symptoms of each case.

Genetic factors or genetic predispositions are crucial because they determine the weakest structural points of the organism. In this way, cerebral inefficiency will tend to manifest first at those structurally weaker points of the organism. However, acquired diseases do not appear solely because of a genetic predisposition. It is necessary for the brain to make many mistakes for them to manifest in the location indicated by the patient's genes. Advances in genetic research allow us to better understand the weak points of the organism and help prevent them from collapsing. But to prevent a genetic predisposition from manifesting as a disease, it will also be necessary to address any affective deficits that may trigger it.

One of the reasons it is difficult to see the relationship between affective deficits and disease in practice is the tremendous resilience of our bodies to abnormalities. Billions of years of evolution have equipped us with an organism capable of withstanding significant trials. Therefore, often, only after several years, an affective deficit manifests as a disease, making it extremely challenging to establish a causal relationship between them. However, as I mentioned earlier, this situation seems to be changing because, since the Industrial Revolution, genetic selection is disappearing. Each new generation of industrial humans incorporates weak, if not harmful, genetic variations that do not disappear because conditions of extreme abundance allow their reproduction, becoming part of the genetic heritage of the population. The result is that each new human generation is genetically weaker than the previous one. Therefore, it is expected that the time needed for an affective deficit to manifest as a disease will shorten in the coming generations, and its impact on human health will become more evident.

But not only do genetic factors play a role in identifying weak points in the organism. Another significant group of factors are cultural ones. Culture, or the information stored physically in the brain, constitutes the living "software" of the organism and determines a significant portion of its behavioral orientation. Cultural information predisposes the brain to selectively attend to some stimuli over others, to give more importance to certain things than others. Therefore, we can also speak of a cultural predisposition to certain diseases. An example will help illustrate how cultural predisposition works. Suppose a person places enormous importance on their external image, on how others perceive them. Their brain will be programmed to primarily focus on anything that could affect their external image. Therefore, the brain will tend to neglect the functions of internal organs, which do not have an external manifestation. The result will be that if this person experiences chronic affective deficit, they may develop a disease that delays its external manifestation as much as possible, such as a heart attack or cancer, for instance. In other words, cultural factors have determined or limited the location of a disease. Another very common example is when a person has a high level of responsibility towards others and, therefore, cannot afford the "luxury" of being ill. For many years, they may not show any symptoms or weakness. But one day, inexplicably, they fall seriously and irreversibly ill. The incidence of cultural factors, such as external image or responsibility towards others, is still poorly understood, and further research is needed.

Lastly, environmental factors, such as geographical and socio-economic ones, also play a crucial role. Diseases are distributed heterogeneously based on the habitat and socio-economic status of patients. It is well known that factors like nutrition, sunlight, atmospheric pollution, humidity, and many environmental factors determine the manifestation of a disease. Likewise, economic and social status determines access to certain resources that impact the occurrence of specific diseases. This group of factors, along with genetic factors, are the most studied and well-known factors today.

An analogy can help illustrate the impact of these three factors on determining the particular symptomatology of each case. Imagine we place a pressure cooker on a fire, filled with water and its safety valves welded shut. We know that sooner or later it will explode. What caused the explosion? Undoubtedly, it was the excessive heat it received, which raised the internal pressure beyond its resistance limit.

Therefore, the cause of the explosion was the excessive heat received. But where will it explode or in what way will it explode? We can only know that it will explode at its weakest point, which depends on multiple factors. Impurities in the material, the quality of the manufacturing, the strength of the welds, etc., are factors that will decide where, when, and how the pressure cooker will burst.

What I am suggesting is that the non-infectious disease of an organism, resulting from cerebral inefficiency, is like the pressure cooker bursting. The disease caused by an affective deficit manifests in the weakest point of the organism, determined by the simultaneous interplay of multiple genetic, cultural, and environmental factors. In general, we can distinguish four major classes of neurological symptom manifestations:

1) Psychological disorders: depression, anxiety, phobia, obsession, etc.
2) Risk behaviors: reckless driving, drug use, obesity, etc.
3) Developmental deficits: academic or employment failure, reproductive issues, etc.
4) Social violence: murder, abuse, rape, robbery, etc.

Genetic, cultural, and environmental factors determine the specific symptoms in each individual patient. But all four types of anomalies stem from a poor and inefficient performance of the brain in processing vital information necessary for achieving the survival and health of the organism.

In general, cerebral inefficiency resulting from systematic affective deficit leads to some disorder in brain function that manifests as one of the mentioned anomalies. These, in turn, eventually lead to the development of some somatic disease. However, I believe there is no reason why cerebral inefficiency cannot manifest directly and exclusively in a somatic disease, although it is not typically the case.

Finally, I think there are reasons to believe that even in infectious diseases, there is an influence of affective deficit. Although these diseases are caused by microbial agents, it is known that the organism has defense mechanisms against them. And from what we know about the brain, the immunological capacity of an organism is directly affected by brain function and, more importantly, indirectly through engaging in risk behaviors that promote disease spread. Therefore, an organism's vulnerability to microbial attacks can also be related to the presence of affective deficit (Graham et al., 2006; Kiecolt-Glaser et al. 2002).

In summary, many non-communicable diseases and most behavioral disorders are caused by affective predation relationships in the affected individuals, and multiple factors (genetic, cultural, and environmental) determine the way the disease manifests and its symptomatology.

The treatment of the disease

The diagnosis of non-communicable diseases should include an analysis of the affected individual's affective relationships in order to determine the presence of a probable affective deficit. The consequence of all this is that no treatment can be effective if it does not directly address the predation relationships that cause the disease. This also explains why Medicine and Clinical Psychology have so many difficulties in making progress in the treatment of these diseases. Taking action on the patient makes no sense and has no beneficial effect if there is no substantial change in the predation relationships affecting them.

Therefore, in addition to providing the appropriate treatment for the symptoms, efforts should be made to guide the patient in resolving specific deficitary relationships that are at the root of the disease. If the cause of the disease is not also addressed, it is expected that the same or another disease will manifest again over time, and so on. On the contrary, if the disease is treated along with affective relationships, meaning that the patient changes the affective relationships that were harming them, then the chances of recovery are significant (depending on the medical treatment that can be provided).

However, once a disease has occurred, it is a mistake to think that it can be cured solely by eliminating the affective deficit that caused it. Although the body has some capacity for self-recovery, a disease is usually, in most cases, an irreversible degradation that can only be recovered through appropriate external medical intervention. In other words, the diagnosis of an affective deficit and its reduction or elimination only has preventive effects on the disease, not curative ones.

Sometimes, when a person becomes seriously ill, they make a significant change in their affective relationships, often reducing or eliminating existing affective deficits. The patient may not be aware of it, but the result is usually a very satisfactory recovery and a favorable prognosis. Many changes in affective relationships occur as a consequence of an illness.

Biopsychology can play an important role in guiding and advising the patient to ensure that this process does not happen only sporadically and randomly. It is in this sense that I believe Biopsychology can assist Medicine in its ultimate goal of achieving the well-being and health of individuals.

Affective relationships in today's society

The affective parasitism that I have described as the underlying cause of the vast majority of non-communicable diseases we suffer from originates in the emotional relationships we maintain with the people closest to us.

In this final section, I have attempted to outline some prototypical examples of this type of relationship in modern societies. All of these examples have been drawn from my years of research at the Orexis clinic, which was run by my mentor, Pilar Gonzalez.

Human isolation

As industrial society advances, emotional relationships become more violent and difficult. One of the reasons for this lies in the progressive isolation of the human species from other species (Wilson, 1984). In an agricultural society, humans were in constant contact with other species, while their contact with other humans was much scarcer (Bell, 1973). Whether it was with domesticated animals, wild animals in hunting, or species of wild or cultivated plants, humans obtained a source of affection from the species they coexisted with (Katcher & Beck, 1987; Melson, 2001). We have seen that affection can be obtained both by receiving positive affection (support, companionship, affection, help, etc.) and by dissipating negative affection (insults, bad moods, scorn, harm, etc.). The agricultural human could "unload" their tension on domesticated animals, wild animals, and plants to a much greater extent than the modern industrial human.

Currently, the only source of affection for the vast majority of industrial humans is other human beings with whom they interact (Putnam, 2000). Therefore, it is logical that emotional relationships are becoming increasingly tense and violent among humans since there is no other alternative to obtain the affection needed for survival.

Killing an animal, for example, is a way to release the aggression (negative affection) that we carry inside. In an agricultural society, a person could, with many differences depending on their social situation, kill many animals with their own hands. Currently, an industrial human hardly ever kills a single animal in their lifetime.

The continued presence of domesticated animals can only be understood from this perspective. The daily contact with an animal can provide an alternative or supplementary source of affection received from fellow humans. Thus, many domesticated animals endure the company of their owners, meaning their whims, impulses, tantrums, indifference, mistreatment, etc. This way, humans can release some of their aggression outside the human social sphere. Of course, the opposite case can occur where the animal receives affection from the human, but in general, it is the other way around.

In summary, the current life of industrial humans forces them to obtain emotional resources exclusively from their fellow humans, making emotional relationships much harsher and violent. Social isolation is the main cause of parasitism and predation in human emotional relationships. People who, for one reason or another, do not have a wide range of social relationships have no choice but to obtain the affection they need from the few people they interact with, thus becoming parasites. When isolation is severe, a person can then become an emotional predator.

The extinction of the family

Perhaps what currently contributes to a significant deterioration of affective relationships is the extinction of the family. The family is an extraordinarily ancient social survival strategy. Many species have developed this strategy, and humans have inherited it from their primate ancestors (Goodall, 1986). Therefore, the family has existed for many millions of years.

The industrial revolution that began in the 18th century with the advent of the steam engine represents an ecological change of unimaginable and unprecedented dimensions (Hobsbawm, 1962). For starters, it seems clear that one of its countless effects is the disappearance of the family as a survival strategy. Until the 19th century, cultural development was directed towards consolidating family structures because human survival was not possible without the resource of the family (Laslett, 1983). Anyone who lost their family also lost their chances of survival. Thus, the family was the central core of our entire culture.

The problem arises when suddenly and dramatically the family ceases to be necessary for survival. This abrupt change has not been a gradual process that has taken place over thousands of years; rather, it has occurred within two hundred years. Currently, an individual in an industrial society does not need a family to survive. In fact, many individuals survive without any contact with their family or even without a family's existence.

However, our culture (i.e., our software) continues to tell us that we should form families, that the family is the most important thing in a person's life, the most suitable, the best. This forms the basis for many pathologies that develop within family relationships.

First, if we examine traditional romantic relationships, we can see how everything tends to lead to one person parasitizing the other over time, meaning that the relationship becomes imbalanced (Fromm, 1956; Beattie, 1987). The reason for this can be understood as follows: in the agrarian society prior to the 19th century, an unwritten law

229

existed stating that if one spouse died, the other spouse was likely to follow shortly after, and vice versa. In other words, there was complete dependence between marriage partners for their survival (Goode, 1963; Laslett, 1972). This resulted in only those couples who maintained a mutually beneficial relationship, taking care of each other, surviving. In all couples where parasitism, a permanent imbalance in the support balance between partners, occurred, they did not survive.

Today, this law has been abolished, and if one of the two members of a couple dies, there is no direct threat to the other's life. They can survive on their own, find another partner, or do whatever they see fit. Therefore, there is no risk for the parasite, creating optimal conditions for its development. Thus, the increase in separations and divorces is unstoppable, and in all cases, the same thing is observed: a parasite and a host. It is the host who decides to separate, and it is the parasite who fights to prevent the separation. Of course, there are cases where this situation does not seem to apply, but it is often well camouflaged. If examined carefully and deeply, we will discover the parasite and the host.

It is this situation, i.e., that the advent of the industrial revolution has favored an optimal situation for the development of the parasite in romantic relationships, that leads us to the rapid extinction of the family. I believe I am not mistaken in stating that the pathology associated with this process of family extinction is the greatest epidemic of the 20th century and probably even the 21st century.

The romantic relationship: the money-affection imbalance

We have stated that the situation in advanced industrial societies leads to couples becoming imbalanced sooner or later. Although there are many different forms of imbalance, depending on each particular case, experience teaches us that there is a very common general pattern today in which women tend to fare worse (Kiecolt-Glaser & Newton, 2001).

Human development in agrarian societies led to specialization within couples in obtaining resources from the external environment. Men specialized in obtaining economic resources (food, defense, territory, etc.), while women specialized in obtaining affective resources (caring for animals and children) (Parsons, 1955). This way, the exchange could be balanced because the effort the man had to make to obtain food was compensated for by the care and attention he received from his wife. She received food for herself and her children in exchange for significant effort in taking care of her husband (even accepting mistreatment).

However, the industrial revolution drastically changes access to economic resources. Over time, the acquisition of food and other economic resources becomes easier, significantly reducing the effort required from men. But concerning affective resources, the situation does not change significantly, and women must continue to make significant efforts to obtain them (Hochschild, 1989).

In this situation, men continue to demand affective resources from their wives, and women find themselves in a clear state of imbalance. This is why it is generally women who decide to break up with their partners, as the situation is clearly unfavorable to them.

The parent-child relationship: education

The fundamental relationship for a person's development is their relationship with their parents or their upbringing (Bowlby, 1988). Success or failure in survival is primarily determined in this relationship. When we are born, we are incapable of obtaining resources by ourselves, meaning we are unable to survive. Therefore, the sole purpose of the parent-child relationship is to ensure that children become capable of surviving on their own and obtaining the resources they need (Erikson, 1950). To achieve this, parents must invest a tremendous amount of resources in their children over many years.

Indeed, it's essential to understand that everything a child lacks to survive must be provided by their parents, as anyone else may seek to take advantage of the child's vulnerability. Therefore, educational success depends solely on the resources parents possess (Coleman, 1988). We are referring to all types of resources, and affectionate resources are absolutely indispensable (Maslow, 1943).

This implies debunking a widespread misconception. In general, parents believe that being with their children, taking care of them, giving them affection, etc., is what matters most. This is a serious mistake because if parents do not have the resources for their children, being with them harms them. In fact, unfortunately, many parents are with their children due to difficulties in obtaining resources from their social environment and end up "consuming" their own children.

Therefore, the primary task of parents is to "go out hunting," which means going out to obtain as many resources as possible. Only in this way, when they return home with their children, can they offer them what they need to survive and grow. It doesn't matter if a parent spends little time with their children; what matters is that they strive to obtain as many resources as possible. Of course, if a parent is highly successful in obtaining resources, they will spend more time with their children. However, if not, it's much better that they go out and continue the struggle rather than staying with them with an empty basket.

What does education consist of? Knowing how to survive means knowing how to fight, and therefore, education involves teaching how to fight. Education has no other purpose than to provide strategies, mechanisms, skills, knowledge, attitudes, etc., that provide the highest possible success in the struggle for life. But how does one learn to fight? Simply by fighting. In other words, education means progressively training our child in the struggle for life, and the key or success of education lies in getting the progression of this training right.

There are two dangers. Either the progression is too slow (overprotection), in which case the child does not develop enough weapons for the fight, or it is too fast (underprotection), in which case the child will be irreversibly harmed in the process. Both extremes lead to the same result: failure in the struggle for existence.

Of course, the ideal or middle point of progression is unattainable in practice because no one has enough resources to maintain it throughout the process. Either we lean towards underprotection, demanding more than they can absorb, or towards overprotection, demanding less than they are capable of.

In any case, it doesn't seem like bad advice to parents to try to experience both types of situations alternately, as this will keep them as close as possible to the ideal point. In other words, if at a certain point they realize they are overprotecting their child, then try to increase the level of demand. If, on the contrary, they feel they are underprotecting their child, then try to reduce the demand. Of course, in the end, this will depend on the resources parents have, as if they do not have sufficient resources, they will not be able to effectively regulate the progression of the educational process and will lean towards one extreme or the other.

The parent-child relationship: old age

Another serious source of pathology today is the crisis in the parent-child relationship in old age. To understand it, we must go back to this momentous fact: the industrial revolution.

As an individual approaches old age, they gradually lose their ability to obtain resources for themselves. In fact, old age becomes a period of weakness, much like childhood. However, in an agricultural society, this relationship could still be positive for the children because their parents possessed essential knowledge for survival. The children survived under almost exactly the same conditions as their parents, and therefore, their parents, with their greater experience, had very valuable knowledge for them. In exchange for this knowledge, the children provided for their parents (economically and emotionally). In a very challenging ecosystem (such as the agricultural one), possessing good knowledge meant an advantage in the struggle for survival. Hence, in cultural evolution, a powerful value system has developed with the ultimate goal of promoting the care and appreciation of parents in their old age.

However, the industrial revolution disrupts the situation, at least until the present moment. It is a revolution of such magnitude that over the last one hundred and fifty years, each generation has had to face an ecosystem completely different from the one the previous generation experienced. This implies that the knowledge held by older people is of little use to the next generation. The means of production have changed, social values have changed, customs, communication systems, and so on. Therefore, elderly parents cannot provide any benefit that could compensate for the effort and cost it takes for their children to continue providing them with economic and, fundamentally, emotional resources.

Therefore, unless for some reason an older parent can still obtain resources for themselves, the widespread caregiving by children to their elderly parents involves an essential loss of resources for their own survival and that of their offspring. Children, compelled by the enormous cultural machinery, feel obligated to provide assistance and affection to their elderly parents without receiving a balanced compensation. The result is a gradual weakening of themselves and, consequently, their own children, putting the preservation of their lineage in jeopardy. We can say that currently, in advanced industrial societies, elderly parents become parasites on their own children, causing them serious harm.

A final consideration

As I have mentioned before, I am convinced that the biological evolution of the Human Sciences (of the memes that constitute them) inevitably tends towards their placement within the framework of Biology. It is in this context that we will begin to understand the truth about human beings, about ourselves, without utopian veils and with the necessary tools that allow us to face the countless challenges that the human species faces today. Personally, I am pessimistic about the survival of the human species, but I am convinced that if there is any possibility that it will not go extinct, one of the factors that can contribute to it will be real knowledge, without fear or prejudice, of our own nature, naked before ourselves and unafraid of our truth. I say this because I believe that nothing is more unbearable for a scientific mind than the fear of the truth. Personally, I have no interest in the ideas presented here as they deeply displease me. I am willing to accept any other explanation as long as it is supported by known facts. Moreover, it would be a source of immense joy for me to find out that this explanation is incorrect and does not fit reality, but I cannot do this by violating the principles of scientific thinking. I know that many people may be able to do it, and I envy them for it. In any case, we are all determined to continue our path, and there is nothing we can do to change it. I feel that I am not culturally prepared to coexist with this view of myself and that I am condemned to suffer the consequences of such dysfunctionality. I hope that future generations will adapt better.

Esteve Barrull, 2007

Conclusions of the first reader

The thermodynamic theory of life has been developing since Schrödinger published his famous *What is Life*, and it has generated significant scientific literature. On the other hand, information theory has also offered a prolific literature in recent years in the study of the nervous system, which will undoubtedly continue to grow in the coming years. The perspective expressed in this book provides a comprehensive view, combining thermodynamics with information theory, furthering our understanding of the human being based on the theory of dissipative structures.

You have probably noticed that throughout the book, the behavior of living beings is described in terms of entropy and at the same time in terms of energy. Energy and entropy are intimately related. Whenever we say that a living being reduces its entropy, we are also saying that it is obtaining free energy, useful energy that it can use to do work, in other words, it is obtaining energy to be able to act. Free (useful) energy and entropy are two sides of the same scale. When free energy increases, entropy decreases, and conversely, when entropy rises, free energy decreases. All phenomena of life are related to this balance. Life is maintained because we struggle, work, and make an effort to get rid of entropy and obtain free energy that allows us to continue acting, working, and struggling. If one stops striving to obtain free energy or is unable to dissipate enough entropy, life goes out forever.

However, the second law of thermodynamics forces us to understand that none of this is free. Obtaining free energy, reducing entropy, cannot be done without consequences. Whenever free energy is obtained, entropy is generated in the environment. Whenever we reduce our entropy, we take free energy away from the environment, and in doing so, we degrade it, impoverish it, disorder it, and bring it closer to thermal death, which is thermodynamic equilibrium. Even more so, the second law requires that the degradation generated in the environment is greater than the free energy or the reduction of the entropy of the living system itself. In other words, life, living beings, exist because they live in an environment rich in free energy, and they exist as long as they can continue to obtain sufficient free energy from it and can continue to degrade it, ultimately destroying it.

This point I just mentioned is exceptionally important and leads us to what I believe is the central theme of this book: the struggle for entropy. I find my father's perspective on the struggle for life very interesting. *The struggle for entropy*, as he used to say, is ultimately the struggle for free energy. It's obvious and evident that living beings compete and fight for resources, which are ultimately various forms of free energy. But expressing it in terms of entropy is indeed interesting. Because, as I've mentioned, obtaining free energy, getting the resources necessary for life, is not free of consequences, and it comes at the expense of our environment, which consists of other living beings that we degrade and destroy to sustain ourselves. I think it's really important to keep this in mind in our lives. Everything we have, we take from others. Everything we are is thanks to the environment we've stripped it from. This is the cruelty of life; our well-being relies on thousands of other lives degraded and destroyed. So, we must not forget that the struggle for resources, the struggle to obtain free energy, is simultaneously a struggle to generate entropy, to degrade the other, because the degradation of the environment is an indispensable condition for the existence of life.

From this perspective, we can see that diseases caused by microorganisms occur within this struggle for entropy. Viruses and bacteria that make us sick feed on our free energy, and that's why they generate entropy in us. All the free energy they obtain is taken from their environment, which is us. And in doing so, they disorder us, degrade us, make us sick. They take away our free energy and generate entropy to the point that if our body cannot overcome them, they ultimately lead to death.

However, the struggle for entropy is not only related to thermodynamics but also extends to the realm of information. Uncertainty and entropy are closely related. The mathematical identity between them, as well as the experimental link shown through energy, suggests that it's not unreasonable to think that uncertainty and entropy are two sides of the same coin. In other words, an information process, a reduction of uncertainty, consists of a physical process of reducing entropy, and vice versa.

By understanding this relationship, especially after discovering Norwich's law of perception, my father realized that the nervous system functions as a dissipative structure of uncertainty. The nervous system acquires uncertainty through perception and dissipates it through action.

When we perceive the world through our senses, our nervous system becomes disordered, acquires entropy, and generates electromagnetic signals proportional to the perceived uncertainty. These signals circulate through the nervous system and eventually reach the muscles or ganglia responsible for our actions. Through movement, organisms can change their position in the world, acquiring information and reducing the perceived uncertainty.

In this sense, we observe the perception-action cycle, which is activated by the perception of uncertainty in the environment and continues until the organism acquires the necessary information to reduce this uncertainty. In other words, action is a response proportional to the perceived uncertainty, which persists until the organism succeeds in reducing it. The nervous system functions as a dissipative structure of uncertainty, guiding the organism to stay away from environmental uncertainty and thus remain maximally informed in a state of minimal uncertainty.

In this sense, it's consistent to assume that animals' emotions also follow the logic of dissipative structures. If we take the time to observe our emotions, we can quickly understand that emotions are the conscious expression of achieving this objective. We feel positive emotions when, through action, we reduce perceived uncertainty, meaning we obtain information or free energy, and thus, we reduce entropy. Conversely, we experience negative emotions when our action is unable to reduce perceived uncertainty, especially when this uncertainty increases, i.e., when we lose information or free energy, and our entropy increases.

This perspective shows us that the primary function of the nervous system is to obtain information and guide the organism's behavior to minimize uncertainty. This certainly seems coherent with the thermodynamic perspective of life. If we understand that there is an empirical relationship between uncertainty and entropy, it makes sense to imagine that the nervous system functions as an alarm system that responds proportionally to the perceived entropy in the environment and guides the organism's behavior to minimize entropy. Furthermore, this interpretation is consistent with the idea that living beings are thermodynamic systems with very low entropy. The extreme order of life is manifested in the nervous system through information. The nervous system is a complex, highly informed system, whose primary goal is not to process information but to acquire it, to obtain more information to remain in a state of minimal uncertainty.

But as we well know, nothing is free in the game of life. If we accept the empirical correlation between uncertainty and entropy, we must assume that the second law also applies in this case. So when our nervous system succeeds in reducing uncertainty, something must become disordered or degraded in the environment. When we acquire information, our environment must obtain a greater amount of entropy. And since most of the information we possess is acquired from other human beings in our environment, we find ourselves immersed again in the struggle for entropy. In our social relationships, humans strive to obtain affection, which, within this framework, we can define as a flow of energy in the form of work that we do for the benefit of others. When we help others, we make an effort, perform work, expend free energy, and therefore, our entropy increases.

It's this interpretation of affection that leads us to understand that the affective deficit can lead to acquired diseases that are not caused by microorganisms. Non-communicable diseases, those that apparently do not originate from other organisms, can have their origins in the affective deficit. In other words, they do have their origins in other organisms, although they are not precisely microscopic; they are other human beings in our environment. When a person does not receive enough affection or when they give much more affection than they receive, they may end up developing non-communicable diseases. This is because a person's work capacity, the amount of free energy available, can be overwhelmed by their workload when they have to take care of themselves and others. So when a person has an affective parasite in their environment, constantly demanding help and taking advantage of their free energy, they end up developing some type of non-communicable disease if they do not have other social relationships that provide enough support to compensate for it. As we've mentioned, our nervous system must stay in a state of minimal entropy, and when it fails to do so, the nervous system starts to malfunction. If our nervous system suffers from an affective deficit, and therefore does not have enough free energy or information to maintain a state of minimal uncertainty, it will eventually suffer from cerebral inefficiency and therefore develop diseases related to the malfunction of the nervous system.

The struggle for entropy is, therefore, the primary cause of the vast majority of acquired diseases. Both communicable and non-communicable diseases have their origins in the struggle with other living beings in the environment. As we saw at the beginning, all the free energy we obtain comes from our environment, which we degrade and ultimately destroy in order to live. So the struggle for affection, for free energy, or for the information provided by other human beings, is also a struggle for entropy, to keep ourselves informed, away from uncertainty in a state of minimal entropy. It is true that this struggle is much more sophisticated than what we can observe when we see lionesses hunting their prey. But in reality, even though we dress it up in luxurious clothing, our struggle for affection is just as cruel and produces as much death as what we observe in nature documentaries.

This is why this book is important. It's important for us to understand that life is a fierce and deadly struggle for entropy that extends to all levels of our experience. Although it may seem like we live in a safe environment, surrounded by concrete and friendly neighbors, the struggle continues, it doesn't stop, and if we want to stay alive, we must pay special attention to the flows of free energy, affection, or information around us.

Elià Barrull Prat, 2023

Bibliographic references

Adams, F. C., & Laughlin, G. (1997). A Dying Universe: The Long-term Fate and Evolution of Astrophysical Objects. *Reviews of Modern Physics*, 69(2), 337-372.

American Psychiatric Association. (2000). *Diagnostic and Statistical Manual of Mental Disorders* (4th ed., text rev.). Washington, DC: Author.

Anderson, G. F., & Poullier, J. P. (1999). Health spending, access, and outcomes: trends in industrialized countries. *Health Affairs*, 18(3), 178-192.

Anderson J. R. (1996). ACT: a simple theory of complex cognition. *Am. Psychol.* 51, 355–365. 10.1037/0003-066x.51.4.355

Antonovsky, A. (1987). *Unraveling the mystery of health: How people manage stress and stay well*. Jossey-Bass.

Aristóteles. (2002). *Nicomachean Ethics*. Cambridge University Press.

Atkins, P. (2007). *Four Laws That Drive the Universe*. Oxford University Press.

Ball, P. (1999). *The Self-Made Tapestry: Pattern Formation in Nature*. Oxford University Press.

Barrera, M., Jr. (2000). Social support research in community psychology. In J. Rappaport & E. Seidman (Eds.), *Handbook of community psychology* (pp. 215-245). Springer US.

Barrull, E. (1992). *Análisis del comportamiento verbal articulatorio en conversaciones grupales espontáneas*. Biopsychology.org

Barth, J., Schneider, S., & von Känel, R. (2010). Lack of social support in the etiology and the prognosis of coronary heart disease: a systematic review and meta-analysis. *Psychosomatic Medicine*, 72(3), 229-238.

Barton, H., & Husk, K. (2000). Healthy urban planning in practice: experience of European cities. *WHO Regional Publications European Series*, (92), 31-44.

Barton, S. (1994). Chaos, self-organization, and psychology. *American Psychologist*, 49, 5–14. doi:10.1037/0003-066X.49.1.5

Batson, C. D. (1991). *The altruism question: Toward a social-psychological answer*. Hillsdale, NJ: Erlbaum.

Baumeister, R. F., & Leary, M. R. (1995). The Need to Belong: Desire for Interpersonal Attachments as a Fundamental Human Motivation. *Psychological Bulletin*, 117(3), 497-529.

Bear, M. F., Connors, B. W., & Paradiso, M. A. (2001). *Neuroscience: Exploring the Brain*. Baltimore, MD: Lippincott Williams & Wilkins.

Beattie, M. (1987). *Codependent No More: How to Stop Controlling Others and Start Caring for Yourself*. Hazelden Publishing.

Begon, M., Townsend, C. R., & Harper, J. L. (2006). *Ecology: From individuals to ecosystems* (4th ed.). Blackwell Publishing.

Bell, D. (1973). *The Coming of Post-Industrial Society: A Venture in Social Forecasting*. New York: Basic Books.

Bengtson, V. L. (2001). Beyond the nuclear family: The increasing importance of multigenerational bonds. *Journal of Marriage and Family*, 63(1), 1-16.

Berkman, L. F., Glass, T., Brissette, I., & Seeman, T. E. (2000). From social integration to health: Durkheim in the new millennium. *Social Science & Medicine*, 51(6), 843-857.

Berkman, L. F., & Kawachi, I. (Eds.). (2000). *Social epidemiology*. Oxford University Press.

Berkman, L. F., & Syme, S. L. (1979). Social networks, host resistance, and mortality: A nine-year follow-up study of Alameda County residents. *American Journal of Epidemiology*, 109(2), 186-204.

Berlin, I. (1954). *Historical inevitability. Four Essays on Liberty.* Oxford University Press.

Berlin, I. (1958). *Two Concepts of Liberty.* Oxford University Press.

Bergson, H. (1907). *L'évolution créatrice.* Paris: F. Alcan.

Bérut, A., Arakelyan, A., Petrosyan, A., Ciliberto, S., Dillenschneider, R., & Lutz, E. (2012). Experimental verification of Landauer's principle linking information and thermodynamics. *Nature*, 483(7388), 187–190. doi:10.1038/nature10872

Bickerton, D. (1990). *Language and species.* University of Chicago Press.

Brillouin, L. (1953). The negentropy principle of information. *Journal of Applied Physics*, 24(9), 1152-1163. doi: 10.1063/1.1721463.

Brillouin, L. (1959). *La science et la théorie de l'information.* Masson

Brody, G. H. (1998). Sibling relationship quality: Its causes and consequences. *Annual Review of Psychology*, 49(1), 1-24. https://doi.org/10.1146/annurev.psych.49.1.1

Bronfenbrenner, U. (1979). *The Ecology of Human Development: Experiments by Nature and Design.* Cambridge, MA: Harvard University Press.

Brooks, D. R., & Wiley, E. O. (1988). *Evolution as Entropy: Toward a Unified Theory of Biology.* University of Chicago Press.

Brown, D. E. (1991). *Human universals.* Temple University Press.

Boehm, C. (1999). *Hierarchy in the Forest: The Evolution of Egalitarian Behavior*. Harvard University Press.

Boltzmann, L. (1896). *Über die mechanische Bedeutung des zweiten Hauptsatzes der Wärmetheorie*. Verlag von Johann Ambrosius Barth.
Borst, A., & Theunissen, F. E. (1999). Information theory and neural coding. *Nature Neuroscience*, 2, 947–957. doi:10.1038/14731

Boulware, L. E., Liu, Y., Fink, N. E., Coresh, J., Ford, D. E., Klag, M. J., & Powe, N. R. (2005). Temporal relation among depression symptoms, cardiovascular disease events, and mortality in end-stage renal disease: contribution of reverse causality. Clinical Journal of the *American Society of Nephrology,* 16(4), 667-675.

Bowlby, J. (1969). *Attachment and loss: Vol. 1. Attachment.* Basic Books.

Bowlby, J. (1982). Attachment and Loss: Retrospect and prospect. *American Journal of Orthopsychiatry*, 52(4), 664-678.

Bowlby, J. (1988). *A secure base: Parent-child attachment and healthy human development.* Basic Books.

Buchner, E. (1903). *Alkoholische Gärung ohne Hefezellen.* Biochemische Zeitschrift, 1(1), 20-47.

Bunge, M. (1960). *La investigación científica*. Buenos Aires: Editorial Ariel.

Cacioppo, J. T., Hughes, M. E., Waite, L. J., Hawkley, L. C., & Thisted, R. A. (2006). Loneliness as a specific risk factor for depressive symptoms: cross-sectional and longitudinal analyses. *Psychology and Aging*, 21(1), 140-151.

Cacioppo, J. T., & Hawkley, L. C. (2009). Perceived social isolation and cognition. *Trends in Cognitive Sciences*, 13(10), 447-454. https://doi.org/10.1016/j.tics.2009.06.005

Carnot, S. (1824). *Réflexions sur la puissance motrice du feu et sur les machines propres à développer cette puissance*. Bachelier.

Carver, C. S., & Scheier, M. (2002). Control processes and self-organization as complementary principles underlying behavior. *Personality and Social Psychology Review*, 6, 304 –315. doi:10.1207/S15327957PSPR0604_05

Chida, Y., Hamer, M., Wardle, J., & Steptoe, A. (2008). Do stress-related psychosocial factors contribute to cancer incidence and survival?. *Nature Clinical Practice Oncology, 5*(8), 466-475.

Chomsky, N. (1981). *Lectures on government and binding: The Pisa lectures*. Walter de Gruyter.

Clarke-Stewart, K. A. (1988). Parents' effects on children's development: A decade of progress? *Journal of Applied Developmental Psychology*, 9(1), 41-84.

Clausius, R. (1867). *The Mechanical Theory of Heat – with its Applications to the Steam Engine and to Physical Properties of Bodies*. John van Voorst.

Crick, F. (1958). On Protein Synthesis. *In Symposia of the Society for Experimental Biology* (Vol. 12, pp. 138-163).

Crick, F. (1994). *The astonishing hypothesis: The scientific search for the soul*. New York, NY: Scribner.

Cohen, S., & Wills, T. A. (1985). Stress, social support, and the buffering hypothesis. *Psychological Bulletin*, 98(2), 310-357. doi: 10.1037/0033-2909.98.2.310

Cohen, S., Gottlieb, B. H., & Underwood, L. G. (2000). Social relationships and health. In S. Cohen, L. G. Underwood, & B. H. Gottlieb (Eds.), *Social support measurement and intervention: A guide for health and social scientists* (pp. 3-25). Oxford University Press.

Coleman, J. S. (1988). Social Capital in the Creation of Human Capital. *American Journal of Sociology*, 94, S95-S120.

Collell G, Fauquet J. (2015). Brain activity and cognition: a connection from thermodynamics and information theory. *Front Psychol*. 6:818. doi: 10.3389/fpsyg.2015.00818.

Combes, C. (2001). *Parasitism: The ecology and evolution of intimate interactions*. Chicago: University of Chicago Press.

Copérnico, N. (1543). *De revolutionibus orbium coelestium* (Sobre las revoluciones de las esferas celestes). Johannes Petreius.

Cox, M. P. (2005). *Understanding Families: A Global Introduction*. Thousand Oaks, CA: Sage Publications.

Cutler, D., & Miller, G. (2005). The role of public health improvements in health advances: The twentieth-century United States. *Demography*, 42(1), 1-22.

Damasio, A. (1994). *Descartes' Error: Emotion, Reason, and the Human Brain*. G.P. Putnam's Sons.

Darwin, C. (1859). *On the Origin of Species by Means of Natural Selection, or the Preservation of Favored Races in the Struggle for Life*. John Murray.

Darwin, C. (1871). *The Descent of Man and Selection in Relation to Sex*. John Murray.

Darwin, C. (1872). *The expression of the emotions in man and animals*. John Murray.

Davidson, R. J., & McEwen, B. S. (2012). Social influences on neuroplasticity: stress and interventions to promote well-being. *Nature Neuroscience*, 15(5), 689-695.

Dawkins, R. (1976). *The Selfish Gene*. Oxford University Press.

Dawkins, R. (1982). *The Extended Phenotype*. Oxford University Press.

Dawkins, R. (2006). *The God Delusion*. Houghton Mifflin.

Decety, J., & Lamm, C. (2006). Human empathy through the lens of social neuroscience. *The Scientific World Journal*, 6, 1146-1163.

Dennett, D. C. (1991). *Consciousness Explained*. Boston, MA: Little, Brown and Company.

Dennett, D. C. (1995). Darwin's Dangerous Idea: Evolution and the Meanings of Life. Simon & Schuster.

DeRosier, M. E., Kupersmidt, J. B., & Patterson, C. J. (1994). Children's academic and behavioral adjustment as a function of the chronicity and proximity of peer rejection. Child Development, 65(6), 1799-1813.

Descartes, R. (1641). *Meditationes de prima philosophia, in qua Dei existentia et animæ immortalitas demonstrantur.* Paris: Michael Soly.

Desimone, R., & Duncan, J. (1995). Neural mechanisms of selective visual attention. Annual Review of Neuroscience, 18, 193-222.

Deutsch, M. (1949). A theory of cooperation and competition. *Human Relations*, 2(1), 129-152.

Diamond, J. (1992). *The Third Chimpanzee: The Evolution and Future of the Human Animal*. New York: HarperCollins Publishers.

Dodge, K. A., Bates, J. E., & Pettit, G. S. (1990). Mechanisms in the cycle of violence. *Science*, 250(4988), 1678-1683.

Dodge, K. A., Coie, J. D., & Lynam, D. R. (2006). Aggression and antisocial behavior in youth. In W. Damon & R. M. Lerner (Series Eds.) & N. Eisenberg (Vol. Ed.), *Handbook of child psychology: Vol. 3. Social, emotional, and personality development* (6th ed., pp. 719-788). John Wiley & Sons.

Doll, R., & Peto, R. (1981). The causes of cancer: Quantitative estimates of avoidable risks of cancer in the United States today. *Journal of the National Cancer Institute*, 66(6), 1191-1308.

Dubos, R. (1959). Mirage of Health: Utopias, Progress, and Biological Change. New York: Harper & Row.

Dubos, R. (1965). Man adapting. Yale University Press.

Dunbar, R. I. M. (1996). *Grooming, Gossip, and the Evolution of Language*. Harvard University Press.

Dunbar, R. I. (1998). The social brain hypothesis. *Evolutionary Anthropology: Issues, News, and Reviews*, 6(5), 178-190.

Durkheim, É. (1893). *The Division of Labor in Society*. The Free Press.

Durkheim, E. (1915). *The elementary forms of religious life*. Free Press.

Dussel, E. (2000). *La filosofía de la liberación ante el nuevo milenio*. Bilbao, España: Desclée de Brouwer.

Egeland, B., & Farber, E. A. (1984). Infant-mother attachment: Factors related to its development and changes over time. *Child Development*, 55(3), 753-771.

Eibl-Eibesfeldt, I. (1986). *Die Biologie des Menschlichen Verhaltens*. Piper

Ekman, P., & Friesen, W. V. (1971). Constants across cultures in the face and emotion. Journal of Personality and Social Psychology, 17(2), 124-129.

Emerson, R. M. (1976). Social exchange theory. *Annual Review of Sociology*, 2, 335-362.

Engel, G. L. (1977). The need for a new medical model: A challenge for biomedicine. *Science*, 196(4286), 129-136. https://doi.org/10.1126/science.847460

Erikson, E. H. (1950). *Childhood and society.* WW Norton & Company.

Ewald, P. W. (1994). *Evolution of infectious disease.* Oxford University Press.

Fagundes, C. P., & Way, B. (2014). Early-life stress and adult inflammation. *Current Directions in Psychological Science,* 23(4), 277-283.

Falk, D. (2004). *Prelinguistic evolution in early hominins: Whence motherese? Behavioral and Brain Sciences*, 27(4), 491-503.

Fechner, G. T. (1850). *Elemente der Psychophysik.* Leipzig, Germany: Breitkopf & Härtel.

Figley, C. R. (Ed.). (1995). *Compassion Fatigue: Coping with Secondary Traumatic Stress Disorder in Those Who Treat the Traumatized.* New York: Brunner/Mazel.

Finkelhor, D., & Ormrod, R. (2001). Child abuse reported to the police. *Juvenile Justice Bulletin.* Washington, D.C.: U.S. Department of Justice, Office of Juvenile Justice and Delinquency Prevention.

Fredrickson, B. L. (2001). The role of positive emotions in positive psychology: The broaden-and-build theory of positive emotions. *American Psychologist*, 56(3), 218-226.

Frenk, J., Bobadilla, J. L., Stern, C., Frejka, T., & Lozano, R. (1991). Elements for a theory of the health transition. *Health Transition Review*, 1(1), 21-38.

Friston, K. (2009). The free-energy principle: A rough guide to the brain? *Trends in Cognitive Sciences*, 13, 293–301. doi:10.1016/j.tics.2009.04.005

Friston, K. (2010). The free-energy principle: A unified brain theory? *Nature Reviews Neuroscience*, 11, 127–138. doi:10.1038/nrn2787

Friston, K., Kilner, J., & Harrison, L. (2006). A free energy principle for the brain. *Journal of Physiology*—Paris, 100, 70 – 87. doi:10.1016/j.jphysparis.2006.10.001

Fromm, E. (1956). *The Art of Loving*. Harper & Row.

Forward, S., & Frazier, D. (1997). *Emotional Blackmail: When the People in Your Life Use Fear, Obligation, and Guilt to Manipulate You*. HarperCollins.

Garbarino, J. (1999). *Lost Boys: Why Our Sons Turn Violent and How We Can Save Them*. Free Press.

Gazzaniga, M. S. (2000). *The mind's past*. University of California Press.

Gazzaniga, M. S. (2005). *The Ethical Brain*. Dana Press.

Gibbs, J. W. (1875 - 1878). *On the Equilibrium of Heterogeneous Substances*. Transactions of the Connecticut Academy of Arts and Sciences,

Gibbs, J. W. (1902). *Elementary Principles in Statistical Mechanics, developed with especial reference to the rational foundation of thermodynamics*. Charles Scribner's Sons.

Gilbert, S. F. (2000). *Developmental Biology* (6th ed.). Sinauer Associates.

Graham, J. E., Christian, L. M., & Kiecolt-Glaser, J. K. (2006). Stress, age, and immune function: Toward a lifespan approach. *Journal of Behavioral Medicine*, 29(4), 389-400.

Godelier, M. (1996). *L'énigme du don*. Fayard.

Goings, T. C., Hidalgo, S. T., & Davidson, C. L. (2019). The impact of social isolation on the risk of adolescent substance use. *Addictive Behaviors*, 90, 79-85.

Goodall, J. (1986). *The chimpanzees of Gombe: Patterns of behavior.* Harvard University Press.

Goode, W. J. (1963). *World Revolution and Family Patterns.* Free Press.

Gould, S. J. (1981). *The Mismeasure of Man.* W.W. Norton & Company.

Gould, S. J. (1989). *Wonderful Life: The Burgess Shale and the Nature of History.* W. W. Norton & Co.

Gouldner, A. W. (1960). The norm of reciprocity: A preliminary statement. *American Sociological Review*, 25(2), 161-178.

Goffman, E. (1959). *The Presentation of Self in Everyday Life.* Anchor Books.

Halfon, N., & Newacheck, P. W. (1999). Prevalence and impact of parent-reported disabling mental health conditions among U.S. children. *Journal of the American Academy of Child & Adolescent Psychiatry*, 38

Hareven, T. K. (2000). *Families, History, and Social Change: Life Course and Cross-Cultural Perspectives.* Boulder, CO: Westview Press.

Harris, S. (2004). *The End of Faith: Religion, Terror, and the Future of Reason.* W. W. Norton & Company.

Harris, S. (2012). *Free will.* New York, NY: Free Press.

Harlow, H. F. (1958). The nature of love. *American Psychologist*, 13(12), 673-685

Harlow, H. F., & Zimmerman, R. R. (1959). Affectional responses in the infant monkey. *Science*, 130(3373), 421-432.

Hauser, M. D., Chomsky, N., & Fitch, W. T. (2002). The faculty of language: What is it, who has it, and how did it evolve?. *Science*, 298(5598), 1569-1579.

Hawkins, J. D., Catalano, R. F., & Miller, J. Y. (1992). Risk and protective factors for alcohol and other drug problems in adolescence and early adulthood: implications for substance abuse prevention. *Psychological Bulletin*, 112(1), 64-105.

Hawkley, L. C., & Cacioppo, J. T. (2010). Loneliness matters: a theoretical and empirical review of consequences and mechanisms. *Annals of Behavioral Medicine*, 40(2), 218-227.

Hawley, A. H. (1950). *Human ecology: A theory of community structure*. New York: Ronald Press.

Hayek, F. A. (1944). *The Road to Serfdom*. University of Chicago Press.

Heise, L. L., Ellsberg, M., & Gottemoeller, M. (1999). Ending violence against women. *Population Reports, Series L*, No. 11. Baltimore, MD: Johns Hopkins University School of Public Health, Population Information Program.

Heidegger, M. (1927). *Sein und Zeit*. Max Niemeyer Verlag.

Hirsh, J.B., Mar, R.A. and Peterson, J.B. (2012). Psychological Entropy: A Framework for Understanding Uncertainty-Related Anxiety. *Psychological Review.*

Hobsbawm, E. J. (1962). *The Age of Revolution: 1789-1848*. Weidenfeld & Nicolson.

Hochschild, A. R. (1989). *The Second Shift: Working Parents and the Revolution at Home*. Viking.

Hofstadter, R. (1944). *Social Darwinism in American Thought.* University of Pennsylvania Press.

Hollis, G., Kloos, H., & Van Orden, G. C. (2009). Origins of order in cognitive activity. In S. J. Guastello, M. Koopmans, & D. Pincus (Eds.), *Chaos and complexity in psychology: The theory of nonlinear dynamical systems* (pp. 206 –241). Boston, MA: Cambridge University Press.

Holt-Lunstad, J., Smith, T. B., & Layton, J. B. (2010). Social relationships and mortality risk: a meta-analytic review. *PLoS Medicine,* 7(7), e1000316.

House, J. S., Landis, K. R., & Umberson, D. (1988). Social relationships and health. *Science,* 241(4865), 540-545. doi:10.1126/science.3399889

Humboldt, A. von (1849). *Cosmos: A sketch of a physical description of the universe.* Harper.

Hutchinson, G. E. (1957). Concluding remarks. *Cold Spring Harbor Symposia on Quantitative Biology*, 22, 415-427.

Ingersoll-Dayton, B., Morgan, D., & Antonucci, T. (1997). The effects of positive and negative social exchanges on aging adults. *The Journals of Gerontology Series B: Psychological Sciences and Social Sciences*, 52(4), S190-S199.

Janis, I. L. (1972). *Victims of groupthink: A psychological study of foreign-policy decisions and fiascoes.* Houghton Mifflin.

Jansson, S.-M., Harjutsalo, V., Forsblom, C., Tuomi, T., & Groop, P.-H. (2012). Predictors of decline in renal function in patients with type 2 diabetes. *Diabetes Care*, 35(11), 2313–2318. doi: 10.2337/dc12-0167

Jaynes, E. T. (1957). *Information Theory and Statistical Mechanics.* The Physical Review.

Johanson, D., & Edey, M. A. (1990). *Lucy: The beginnings of humankind.* Simon & Schuster.

Johanson, D., & Edgar, B. (2006). *From Lucy to language.* Simon & Schuster.

Kandel, E. R., Schwartz, J. H., & Jessell, T. M. (2000). *Principles of Neural Science* (4th ed.). McGraw-Hill.

Kandel, E. R. (2001). The Molecular Biology of Memory Storage: A Dialogue Between Genes and Synapses. *Science*, 294(5544), 1030-1038.

Kass, L. R. (1985). *Toward a More Natural Science: Biology and Human Affairs.* Nueva York: Free Press.

Kauffman, S. A. (1993). *The origins of order: Self organization and selection in evolution.* New York, NY: Oxford University Press.

Kauffman, S. A. (2000). *Investigations.* Oxford University Press.

Kawachi, I., & Berkman, L. F. (2001). Social ties and mental health. *Journal of Urban Health*, 78(3), 458-467.

Kelley, H. H., & Thibaut, J. W. (1978). *Interpersonal relations: A theory of interdependence.* Wiley.

Kendler, K. S., Prescott, C. A., Myers, J., & Neale, M. C. (2001). The structure of genetic and environmental risk factors for common psychiatric and substance use disorders in men and women. *Archives of General Psychiatry*, 60(9), 929-937.

Kahn, R. L., & Antonucci, T. C. (1980). Convoys over the life course: Attachment, roles, and social support. In P. B. Baltes & O. G. Brim (Eds.), *Life-span development and behavior* (Vol. 3, pp. 253–286). Academic Press.

Kiecolt-Glaser, J. K., & Newton, T. L. (2001). Marriage and health: his and hers. *Psychological Bulletin*, 127(4), 472-503.

Kiecolt-Glaser, J. K., McGuire, L., Robles, T. F., & Glaser, R. (2002). Emotions, morbidity, and mortality: new perspectives from psychoneuroimmunology. *Annual Review of Psychology*, 53, 83-107.

Knol, M. J., Twisk, J. W., Beekman, A. T., Heine, R. J., Snoek, F. J., & Pouwer, F. (2006). Depression as a risk factor for the onset of type 2 diabetes mellitus. A meta-analysis. *Diabetologia*, 49(5), 837-845.

Krebs, J. R., & Davies, N. B. (1993). *An introduction to behavioural ecology* (3rd ed.). Oxford: Blackwell Scientific Publications.

Kropotkin, P. (1902). *Mutual Aid: A Factor of Evolution*. London: William Heinemann.

Koffka, K., (1935). *Principles of gestalt psychology*. Brace and Company.

Köhler, W. (1947). *Gestalt Psychology*. Liveright Publishing Corporation.

Kolb, B., & Whishaw, I. Q. (2009). *Fundamentals of Human Neuropsychology* (6th ed.). Worth Publishers.

La Cerra P. (2003). The first law of psychology is the second law of thermodynamics: the energetic evolutionary model of the mind and the generation of human psychological phenomena. *Hum. Nat. Rev.* 3, 440–447.

Lagarde, E., Chastang, J. F., Gueguen, A., Coeuret-Pellicer, M., Chiron, M., & Lafont, S. (2004). Emotional stress and traffic accidents: the impact of separation and divorce. *Epidemiology*, 15(6), 762-766.

Lakatos, I. (1970). Falsification and the Methodology of Scientific Research Programmes. In I. Lakatos & A. Musgrave (Eds.), *Criticism and the Growth of Knowledge* (pp. 91-195). Cambridge University Press.

Landauer, R. (1961). Irreversibility and heat generation in the computing process. *IBM Journal of Research and Development*, 5(3), 183-191.

Laske, R. E. (1997). Intergenerational relationships and their consequences for old age. In V. L. Bengtson (Ed.), *Adulthood and ageing: Research on continuities and discontinuities* (pp. 125-149). Springer Publishing Company.

Laslett, P. (1972). *The World We Have Lost: England Before the Industrial Age.* Charles Scribner's Sons.

Laslett, P. (1983). *Family and kinship in Europe.* In Family forms in historic Europe (pp. 1-89). Cambridge University Press.

Laughlin, S. B. (1989). The role of sensory adaptation in the retina. *Journal of Experimental Biology*, 146, 39-62.

Le Chatelier, M.H. (1888). *Recherches expérimentales et théoriques sur les équilibres chimiques.* Libraire des corps nationaux des ponts et chaussées des mines et des télégraphes.

Leakey, R. E. (1981). *The Making of Mankind.* Sherma B. V.

Leakey, R. E. (1994). *The origin of humankind.* Perseus Books Group.

Leakey, R. E., & Walker, A. C. (1997). *The Nariokotome Homo erectus Skeleton.* Harvard University Press.

LeDoux, J. (1996). *The Emotional Brain: The Mysterious Underpinnings of Emotional Life.* New York: Simon & Schuster.

Lenz, E. (1834). Ueber die Bestimmung der Richtung der durch elektrodynamische Vertheilung erregten galvanischen Ströme. *Annalen der Physik und Chemie*, 107(31), 483-494.

Levinas, E. (1961). *Totality and Infinity: An Essay on Exteriority.* Duquesne University Press.

Lewin, K. (1936). *Principles of topological psychology*. McGraw-Hill.

Lewis, C.S. (1943). *The Abolition of Man.* Oxford University Press

Liang, J., Bennett, J. M., Krause, N. M., Chang, M., Lin, H., Chuang, Y., and Wu, S. (1999). Stress, Social Relations, and Old Age Mortality in Taiwan. J *Clin Epidemiol* Vol. 52, No. 10, pp. 983–995

Liang, J., Krause, N. M., & Bennett, J. M. (2001). Social exchange and well-being: Is giving better than receiving? *Psychology and Aging,* 16(3), 511-523.

Lin, N., Woelfel, M. W., & Light, S. C. (1985). The buffering effect of social support subsequent to an important life event. *Journal of Health and Social Behavior*, 26(3), 247-263.

Lovejoy, C. O. (1988). Evolution of Human Walking. *Scientific American*, 259(5), 118-125.

Lupien, S. J., McEwen, B. S., Gunnar, M. R., & Heim, C. (2009). Effects of stress throughout the lifespan on the brain, behaviour and cognition. *Nature Reviews Neuroscience*, 10(6), 434-445.

Lutgendorf, S. K., & Sood, A. K. (2011). Biobehavioral factors and cancer progression: physiological pathways and mechanisms. *Psychosomatic Medicine*, 73(9), 724-730.

Margulis, L. (1970). *Origin of Eukaryotic Cells*. Yale University Press.

Margulis, L., Sagan D., and Eldredge, N. (1995) *What Is Life?*. Simon and Schuster.

Margulis, L., Sagan, D. (1997). *Slanted Truths: Essays on Gaia, Symbiosis, and Evolution*. Copernicus Books.

Marmot, M., & Wilkinson, R. G. (Eds.). (2001). *Social determinants of health*. Oxford University Press.

Maslach, C., & Leiter, M. P. (2008). Early predictors of job burnout and engagement. *Journal of Applied Psychology,* 93(3), 498-512.

Maslow, A. H. (1943). A theory of human motivation. *Psychological Review*, 50(4), 370-396. https://doi.org/10.1037/h0054346

Masten, A. S., & Coatsworth, J. D. (1998). The development of competence in favorable and unfavorable environments: Lessons from research on successful children. *American Psychologist*, 53(2), 205-220.

Mayr, E. (1982). *The Growth of Biological Thought.* The Belknap Press of Harvard University Press.

McEwen, B. S. (1998). Protective and damaging effects of stress mediators. *New England Journal of Medicine,* 338(3), 171-179.

McEwen, B. S. (2007). Physiology and neurobiology of stress and adaptation: Central role of the brain. *Physiological Reviews*, 87(3), 873-904.

McEwen, B. S., & Lasley, E. N. (2003). Allostatic load: when protection gives way to damage. *Advances in Mind-Body Medicine,* 19(1), 28-33.

McEwen, B. S., & Stellar, E. (1993). Stress and the individual: Mechanisms leading to disease. *Archives of Internal Medicine*, 153(18),2093-2101.
https://doi.org/10.1001/archinte.1993.00410180039004

McHenry, H. M. (1994). Tempo and mode in human evolution. *Proceedings of the National Academy of Sciences*, 91(15), 6780-6786.

McHugh, J. E., Lawlor, B. A., & Casey, A. M. (2011). Psychosocial correlates of aspects of sleep quality in community-dwelling Irish older adults. *Aging & Mental Health,* 15(6), 749-755.

McIrvine, E. C. y Tribus. (1971). *Energy and Information.* Scientific American.

Melson, G. F. (2001). *Why the Wild Things Are: Animals in the Lives of Children*. Cambridge, MA: Harvard University Press.

Midgley, M. (1983). *Animals and Why They Matter*. University of Georgia Press.

Mikami, A. Y., & Hinshaw, S. P. (2006). Resilient adolescent adjustment among girls: Buffers of childhood peer rejection and attention-deficit/hyperactivity disorder. Journal of Abnormal Child Psychology, 34(6), 823-837.

Monod, J. (1970). *Le hasard et la nécessité: Essai sur la philosophie naturelle de la biologie moderne*. Éditions du Seuil.

Morowitz, H. J. (1968). *Energy Flow in Biology: Biological Organization as a Problem in Thermal Physics*. Woodbridge, CT: Ox Bow Press.

Nagel, T. (1986). *The view from nowhere*. Oxford University Press.

Neisser, U. (1976). *Cognition and reality: Principles and implications of cognitive psychology*. W. H. Freeman.

Nemenman, I., Bialek, W., & de Ruyter van Steveninck, R. (2004). Entropy and information in neural spike trains: Progress on the sampling problem. *Physical Review*, 69, Article 056111. doi:10.1103/PhysRevE.69.056111

Nesse, R. M., & Williams, G. C. (1994). *Why we get sick: The new science of Darwinian medicine*. Vintage Books.

Newton, I. (1687). *Philosophiæ Naturalis Principia Mathematica*. Londini: Jussu Societatis Regiæ ac Typis Josephi Streater.

Norwich, K. H. (1993). *Information, Sensation and Perception*. Academic Press.

Norwich, K. H. (2001). Determination of Saltiness from the laws of thermodynamics: Estimating the gas constant from psychophysical experiments. *Chemical Senses.* 26, 1015-1022.

Norwich, K. H. (2010). Le Chatelier's principle in sensation and perception: fractal-like enfolding at different scales. *Frontiers in physiology.* Volum 1 Article 17

Odum, E. P. (1971). *Fundamentals of ecology* (3rd ed.). Saunders.

Odum, E. P. (1983). *Basic ecology.* Saunders College Publishing.

Omran, A. R. (1971). The Epidemiologic Transition: A Theory of the Epidemiology of Population Change. *The Milbank Memorial Fund Quarterly*, 49(4), 509-538.

Orth-Gomér, K., Wamala, S. P., Horsten, M., Schenck-Gustafsson, K., Schneiderman, N., & Mittleman, M. A. (2000). Marital stress worsens prognosis in women with coronary heart disease: the Stockholm *Female Coronary Risk Study.* JAMA, 284(23), 3008-3014.

Palmer, S., Vecchio, M., Craig, J. C., Tonelli, M., Johnson, D. W., Nicolucci, A., ... & Strippoli, G. F. (2013). Association between depression and death in people with CKD: a meta-analysis of cohort studies. *American Journal of Kidney Diseases*, 62(3), 493-505.

Parsons, T. (1955). *Family, Socialization and Interaction Process.* Free Press.

Pasteur, L. (1861). Mémoire sur les corpuscules organisés qui existent dans l'atmosphère: examen de la doctrine des générations spontanées. *Annales de chimie et de physique*, 62, 337-361.

Panksepp, J. (1998). *Affective neuroscience: The foundations of human and animal emotions.* Oxford University Press.

Paninski, L. (2003). Estimation of entropy and mutual information. *Neural Computation*, 15, 1191–1253. doi:10.1162/089976603321780272

Pellow, D. N. (2007). *Resisting global toxics: Transnational movements for environmental justice.* MIT Press.

Pereda, E., Quiroga, R., & Bhattacharya, J. (2005). Nonlinear multivariate analysis of neurophysiological signals. *Progress in Neurobiology*, 77, 1–37. doi:10.1016/j.pneurobio.2005.10.003

Piaget, J. (1950). *The psychology of intelligence.* Routledge.

Pianka, E. R. (2000). *Evolutionary ecology* (6th ed.). San Francisco: Benjamin Cummings.

Piferi, R. L., & Lawler, K. A. (2006). Social support and ambulatory blood pressure: An examination of both receiving and giving. *International Journal of Psychophysiology*, 62(2), 328-336.

Pines, A., & Aronson, E. (1988). *Career Burnout: Causes and Cures.* New York: Free Press.

Pines, A. (1996). *Couple burnout: Causes and cures.* Psychology Press.

Pinker, S. (1994). *The language instinct: How the mind creates language.* HarperCollins Publishers.

Pinker, S. (1997). *How the Mind Works.* W.W. Norton & Company.

Pinker, S. (2002). *The Blank Slate: The Modern Denial of Human Nature.* Viking.

Pinquart, M., & Duberstein, P. R. (2010). Associations of social networks with cancer mortality: a meta-analysis. *Critical Reviews in Oncology/Hematology,* 75(2), 122-137.

Pinquart, M., & Sörensen, S. (2003). Differences between caregivers and noncaregivers in psychological health and physical health: A meta-analysis. *Psychology and Aging*, 18(2), 250-267.

Prigogine, I. (1955). *Introduction to Thermodynamics of Irreversible Processes*. Charles C. Thomas Publisher.

Prigogine, I. (1957). *The Molecular Theory of Solutions*. North Holland Publishing Company.

Prigogine, I. (1977). Order through fluctuation: Self-organization and social system. In Jantsch, E. & Waddington, C. H. (Eds.), *Evolution and Consciousness: Human Systems in Transition*. Addison-Wesley.

Prigogine, I. & Nicolis, G. (1977). *Self-Organization in Non-Equilibrium Systems*. Wiley.

Prigogine, I. (1980). *From Being to Becoming: Time and Complexity in the Physical Sciences*. W.H. Freeman and Company.

Prigogine, I. & Stengers, I. (1984). *Order out of Chaos: Man's new dialogue with nature*. Flamingo.

Popper, K. R. (1959). *The Logic of Scientific Discovery*. Basic Books.

Porter, R. (1997). *The Greatest Benefit to Mankind: A Medical History of Humanity*. New York: W. W. Norton & Company.

Poulin, R., & Morand, S. (2000). The diversity of parasites. *The Quarterly Review of Biology*, 75(3), 277-293.

Pouwer, F., Kupper, N., & Adriaanse, M. C. (2010). Does emotional stress cause type 2 diabetes mellitus? A review from the European Depression in Diabetes (EDID) Research Consortium. *Discovery Medicine*, 9(45), 112-118.

Putnam, R. D. (2000). *Bowling Alone: The Collapse and Revival of American Community*. New York: Simon & Schuster.

Rosengren, A., Orth-Gomér, K., & Wilhelmsen, L. (1991). Lack of social support and incidence of coronary heart disease in middle-aged Swedish men. *Psychosomatic Medicine*, 53(4), 357–363. doi: 10.1097/00006842-199107000-00009

Rosengren, A., Hawken, S., Ôunpuu, S., Sliwa, K., Zubaid, M., Almahmeed, W. A., ... & INTERHEART investigators. (2004). Association of psychosocial risk factors with risk of acute myocardial infarction in 11 119 cases and 13 648 controls from 52 countries (the INTERHEART study): case-control study. *The Lancet*, 364(9438), 953-962.

Rothschild, M. (1992). *Bionomics: Economy as Ecosystem*. Henry Holt

Rothman, K. J., Greenland, S., & Lash, T. L. (2008). *Modern epidemiology* (3rd ed.). Lippincott Williams & Wilkins.

Rousseau, J. J. (1762). *Du contrat social ou Principes du droit politique*. Chez Marc-Michel Rey.

Rueger, S. Y., Malecki, C. K., & Demaray, M. K. (2010). Relationship between multiple sources of perceived social support and psychological and academic adjustment in early adolescence: Comparisons across gender. *Journal of Youth and Adolescence*, 39(1), 47-61.

Rutter, M. (1981). Stress, coping and development: Some issues and some questions. *Journal of Child Psychology and Psychiatry,* 22(4), 323-356.

Sabater, P. J. (1978). *El chimpancé y los orígenes de la cultura*. Anthropos.

Santini, Z. I., Koyanagi, A., Tyrovolas, S., Mason, C., & Haro, J. M. (2015). The association between social relationships and depression: a systematic review. *Journal of Affective Disorders*, 175, 53-65.

Sapolsky, R. M. (2004). *Why Zebras Don't Get Ulcers: The Acclaimed Guide to Stress, Stress-Related Diseases, and Coping* (3rd ed.). Henry Holt and Company.

Sartre, J.-P. (1943). *L'Être et le Néant: Essai d'ontologie phénoménologique.* Gallimard.

Schachter, S., & Singer, J. E. (1962). Cognitive, social, and physiological determinants of emotional state. *Psychological Review*, 69(5), 379-399.

Scheff, T. J. (1997). *Emotions, the social bond, and human reality: Part/Whole analysis.* Cambridge University Press.

Schneider, E. D., & Sagan, D. (2005). *Into the Cool: Energy Flow, Thermodynamics, and Life.* University of Chicago Press.

Schrodinger, E. (1944). *What is life.* Cambridge University Press

Schulz, R., & Sherwood, P. R. (2008). Physical and mental health effects of family caregiving. *American Journal of Nursing*, 108(9 Suppl), 23-27.

Scruton, R. (2000). *Animal Rights and Wrongs.* Demos.

Searle, J. R. (1998). *Mind, Language and Society: Philosophy in the Real World.* Basic Books.

Seeman, T. E. (2000). Health promoting effects of friends and family on health outcomes in older adults. *American Journal of Health Promotion*, 14(6), 362-370.

Segerstrom, S. C., & Miller, G. E. (2004). Psychological stress and the human immune system: A meta-analytic study of 30 years of inquiry. *Psychological Bulletin*, 130(4), 601-630.

Segrin, C., & Passalacqua, S. A. (2010). Functions of loneliness, social support, health behaviors, and stress in association with poor health. *Health Communication*, 25(4), 312-322.

Shannon, C. E. y Weaver, W. (1949). *The mathematical theory of communication.* Urbana: The University of Chicago Press.

Sherif, M. (1966). *In common predicament: Social psychology of intergroup conflict and cooperation.* Houghton Mifflin.

Seltzer, M. M., Greenberg, J. S., Floyd, F. J., Pettee, Y., & Hong, J. (2001). Life course impacts of parenting a child with a disability. *American Journal on Mental Retardation,* 106(3), 265-286.

Silverstein, M., & Bengtson, V. L. (1997). Intergenerational solidarity and the structure of adult child-parent relationships in American families. *American Journal of Sociology,* 103(2), 429-460.

Sinha, R. (2008). Chronic stress, drug use, and vulnerability to addiction. *Annals of the New York Academy of Sciences,* 1141(1), 105-130.

Squire, L. R., Bloom, F. E., McConnell, S. K., Roberts, J. L., Spitzer, N. C., & Zigmond, M. J. (2003). *Fundamental neuroscience.* Academic Press.

Spencer, H. (1876). *Principles of sociology* (Vol. 1). D. Appleton and Company.

Sroufe, L. A. (1997). Psychopathology as an outcome of development. *Development and Psychopathology,* 9(2), 251-268.

Stampfer, M. J., Colditz, G. A., Willett, W. C., Speizer, F. E., & Hennekens, C. H. (2000). A prospective study of moderate alcohol consumption and the risk of coronary disease and stroke in women. *The New England Journal of Medicine,* 319(5), 267-273. https://doi.org/10.1056/NEJM198808043190503

Stephen, D. G., Boncoddo, R. A., Magnuson, J. S., & Dixon, J. A. (2009). The dynamics of insight: Mathematical discovery as a phase transition. *Memory & Cognition,* 37, 1132–1149. doi:10.3758/MC.37.8.1132

Stephen, D. G., Dixon, J. A., & Isenhower, R. W. (2009). Dynamics of representational change: Entropy, action, and cognition. *Journal of Experimental Psychology: Human Perception and Performance,* 35, 1811– 1832. doi:10.1037/a0014510

Stevens, S. S. (1957). On the psychophysical law. *Psychological Review*, 64(3), 153-181.

Strong, S., Koberle, R., de Ruyter van Steveninck, R., & Bialek, W. (1998). Entropy and information in neural spike trains. *Physical Review Letters*, 80, 197–200. doi:10.1103/PhysRevLett.80.197

Suldo, S. M., Shaunessy, E., & Hardesty, R. (2008). Relationships among stress, coping, and mental health in high-achieving high school students. *Psychology in the Schools*, 45(4), 273-290.

Szilárd, L. (1929). Über die Entropieverminderung in einem thermodynamischen System bei Eingriffen intelligenter Wesen. *Zeitschrift für Physik*, 53(11-12), 840-856. doi:10.1007/BF01341281.

Taylor, S. E., Klein, L. C., Lewis, B. P., Gruenewald, T. L., Gurung, R. A. R., & Updegraff, J. A. (2000). Biobehavioral responses to stress in females: Tend-and-befriend, not fight-or-flight. *Psychological Review*, 107(3), 411-429. doi: 10.1037/0033-295X.107.3.411

Taylor, S. E., & Stanton, A. L. (2007). Coping resources, coping processes, and mental health. *Annual Review of Clinical Psychology, 3,* 377-401.

Thomas, P. A. (2010). Is it better to give or to receive? Social support and the well-being of older adults. *The Journals of Gerontology Series B: Psychological Sciences and Social Sciences,* 65(3), 351-357.

Tilman, D. (1982). *Resource competition and community structure.* Princeton University Press.

Triandis, H. C., & Suh, E. M. (2002). Cultural influences on personality. *Annual Review of Psychology*, 53, 133-160.

Trivers, R. (1971). The evolution of reciprocal altruism. *The Quarterly Review of Biology*, 46(1), 35-57.

Troisi, A., Di Lorenzo, G., Alcini, S., Nanni, R. C., Di Pasquale, C., & Siracusano, A. (2006). Body dissatisfaction in women with eating disorders: Relationship to early separation anxiety and insecure attachment. *Psychosomatic Medicine*, 68(3), 449-453.

Tolman, E. C. (1932). *Principles of Purposive Behaviorism.* Prentice-Hall, Inc.

Tononi, G., Sporns, O., & Edelman, G. (1994). A measure for brain complexity: Relating functional segregation and integration in the nervous system. *Proceedings of the National Academy of Sciences of the United States of America*, 91, 5033–5037. doi:10.1073/pnas.91.11.5033

Toyabe, S., Sagawa, T., Ueda, M., Muneyuki, E., & Sano, M. (2010). Information heat engine: converting information to energy by feedback control. *Nature Physics*, 6(12), 988–992. doi.org/10.1038/nphys1821

Uchino, B. N. (2004). *Social support and physical health: Understanding the health consequences of relationships.* Yale University Press.

Uchino, B. N. (2006). Social support and health: a review of physiological processes potentially underlying links to disease outcomes. *Journal of Behavioral Medicine*, 29(4), 377-387.

Uchino, B. N. (2009). Understanding the links between social support and physical health: A life-span perspective with emphasis on the separability of perceived and received support. *Perspectives on Psychological Science*, 4(3), 236-255.

Useche, S. A., Ortiz, V. G., & Cendales, B. E. (2018). Stress-related psychosocial factors at work, fatigue, and risky driving behavior in bus rapid transport (BRT) drivers. *Accident Analysis & Prevention*, 106, 191-199.

Umberson, D., & Montez, J. K. (2010). Social relationships and health: a flashpoint for health policy. *Journal of Health and Social Behavior*, 51(1_suppl), S54-S66.

Valkenburg, P. M., & Peter, J. (2007). Online communication and adolescent well-being: Testing the stimulation versus the displacement hypothesis. Journal of Computer-Mediated Communication, 12(4), 1169-1182.

Vallacher, R., Read, S., & Nowak, A. (2002). The dynamical perspective in personality and social psychology. Personality and *Social Psychology Review*, 6, 264 –273. doi:10.1207/S15327957PSPR0604_01

Varpula S., Annila A., Beck C. (2013). Thoughts about thinking: cognition according to the second law of thermodynamics. *Adv. Stud. Biol.* 5, 135–149.

Vesalius, A. (1543). *De humani corporis fabrica*. Basileae: Ex officina Joannis Oporini.

Von Neumann, J. (1932). *Mathematical Foundations of Quantum Mechanics*. Princeton University Press.

Vygotsky, L. S. (1978). *Mind in society: The development of higher psychological processes*. Harvard University Press.

Watson, J. D., & Crick, F. H. C. (1953). Molecular structure of nucleic acids: a structure for deoxyribose nucleic acid. *Nature*, 171(4356), 737-738.

Watson, J. D. & Crick, F. H. C. (1968). *The Double Helix: A Personal Account of the Discovery of the Structure of DNA*. Atheneum Publishers.

Weber, E. H. (1846). *Tastsinn und Gemeingefühl: Eine Untersuchung psychologischer und physiologischer Natur*. Leipzig: Winter.

Wertheimer, M. (1923). Untersuchungen zur Lehre von der Gestalt, II. *Psychologische Forschung*, 4, 301-350.

Willett, W. C. (2002). Balancing life-style and genomics research for disease prevention. *Science*, 296(5568), 695-698. https://doi.org/10.1126/science.1071055

Wilson, E. O. (1975). *Sociobiology: The new synthesis*. Harvard University Press.

Wilson, E. O. (1984). *Biophilia*. Cambridge, MA: Harvard University Press.

Wilson, E. O. (1998). *Consilience: The Unity of Knowledge*. Vintage Books.

Winner, L. (1986). *The Whale and the Reactor: A Search for Limits in an Age of High Technology*. Chicago: The University of Chicago Press.

Wolpert, D. M., Doya, K., & Kawato, M. (2003). A unifying computational framework for motor control and social interaction. *Philosophical Transactions of the Royal Society of London. Series B, Biological Sciences*, 358(1431), 593-602.

Wolpert, L., Beddington, R., Brockes, J., Jessell, T., Lawrence, P., & Meyerowitz, E. (2002). *Principles of Development* (2nd ed.). Oxford University Press.

World Health Organization. (2002). *The World Health Report 2002: Reducing Risks, Promoting Healthy Life*. World Health Organization.

World Health Organization. (2005). *Preventing chronic diseases: a vital investment*. World Health Organization.

Wukmir, V. J. (1960). *Psicología de la orientación vital*. Barcelona: Luis Miracle

Wukmir, V. J. (1967). *Emoción y sufrimiento*. Barcelona: Labor

www.ingramcontent.com/pod-product-compliance
Lightning Source LLC
Chambersburg PA
CBHW072357290526
45794CB00001B/91